Environmental Governance in Taiwan

Three decades of rapid industrialization until the lifting of martial law in 1987, with little or no concern for the environment, have made Taiwan's environmental degradation a serious problem. In the past 20 years, Taiwan has seen a surge of environmental organizations, which to a certain degree have enjoyed a remarkable success in fighting polluting industries or affecting policies on behalf of the environment.

This book aims to analyze environmental governance mechanisms and actors in Taiwan through a multi-disciplinary research approach. Based on extensive and original research, it includes four different case studies, which have all taken place since 2011. It focuses on four major elements of governance – specifically norms, actors, processes and outcomes – to examine Taiwan's national and local environmental governance in the post-2008 period. The book shows how the painful lessons Taiwan has learned throughout its transition should be of interest to other developing countries, illustrating how these positive transformations have managed to bring about a more ecologically friendly mode of economic development.

Demonstrating that the battle to further ecological sustainability is also a battle to further democratization, this book will be of interest to students and scholars of Taiwan Studies, Developmental Studies and Environmental Studies.

Simona A. Grano is Senior Lecturer at the Department of Sinology at the University of Zürich, Switzerland. Her research interests focus on environmental management and politics, state-society relations and judicial sector reforms.

Routledge research on Taiwan

Series Editor: Dafydd Fell
SOAS, UK

The *Routledge research on Taiwan series* seeks to publish quality research on all aspects of Taiwan studies. Taking an interdisciplinary approach, the books will cover topics such as politics, economic development, culture, society, anthropology and history.

This new book series will include the best possible scholarship from the social sciences and the humanities and welcomes submissions from established authors in the field as well as from younger authors. In addition to research monographs and edited volumes general works or textbooks with a broader appeal will be considered.

The series is advised by an international editorial board and edited by Dafydd Fell of the Centre of Taiwan Studies at the School of Oriental and African Studies.

1 **Taiwan, Humanitarianism and Global Governance**
Alain Guilloux

2 **Planning in Taiwan**
Spatial planning in the twenty-first century
Edited by Roger Bristow

3 **Popular Culture in Taiwan**
Charismatic modernity
Edited by Marc L. Moskowitz

4 **Politics of Difference in Taiwan**
Edited by Tak-Wing Ngo and Hong-zen Wang

5 **Taiwanese Identity in the 21st Century**
Domestic, regional and global perspectives
Edited by Gunter Schubert and Jens Damm

6 **Taiwan's Democracy**
Economic and political challenges
Edited by Robert Ash, John Garver and Penelope B. Prime

7 **Taiwan's Economic Transformation**
Leadership, property rights and institutional change
Tai-chun Kuo and Ramon H. Myers

8 **Government and Politics in Taiwan**
Dafydd Fell

9 **Documenting Taiwan on Film**
Issues and methods in new documentaries
Edited by Sylvia Li-chun Lin and Tze-lan D. Sang

10 **Technology Transfer Between the US, China and Taiwan**
Moving knowledge
Edited by Douglas B. Fuller and Murray A. Rubinstein

11 **Migration to and from Taiwan**
Edited by Kuei-fen Chiu, Dafydd Fell and Lin Ping

12 **Political Changes in Taiwan under Ma Ying-jeou**
Continuity, change and future challenges
Edited by Jean-Pierre Cabestan and Jacques deLisle

13 **Border Crossing in Greater China**
Production, community and identity
Edited by Jenn-hwan Wang

14 **Language, Politics and Identity in Taiwan**
Naming China
Hui-Ching Chang and Richard Holt

15 **Place, Identity, and National Imagination in Post-war Taiwan**
Bi-yu Chang

16 **Environmental Governance in Taiwan**
A new generation of activists and stakeholders
Simona A. Grano

Environmental Governance in Taiwan

A new generation of activists and stakeholders

Simona A. Grano

Routledge
Taylor & Francis Group
LONDON AND NEW YORK

First published 2015
by Routledge
2 Park Square, Milton Park, Abingdon, Oxon OX14 4RN

and by Routledge
711 Third Avenue, New York, NY 10017

First issued in paperback 2017

Routledge is an imprint of the Taylor & Francis Group, an informa business

British Library Cataloguing in Publication Data
A catalogue record for this book is available from the British Library

Library of Congress Cataloging-in-Publication Data
Grano, Simona Alba.
Environmental governance in Taiwan : a new generation of activists and stakeholders / Simona A. Grano.
pages cm. – (Routledge research on Taiwan ; 16)
Includes bibliographical references and index.
1. Environmental policy–Taiwan. 2. Taiwan–Environmental conditions.
I. Title.
GE190.T3G73 2015
363.7'05610951249–dc23 2014048226

ISBN 13: 978-1-138-55417-7 (pbk)
ISBN 13: 978-1-138-83140-7 (hbk)

Typeset in Times New Roman
by Wearset Ltd, Boldon, Tyne and Wear

Contents

Figures

Preface

My long journey through the maze of Taiwanese green politics begins in February 2011. While many people mistakenly think of green politics and environmentalism as secondary aspects of a state's main responsibilities, environmental politics reveal a great deal about how governments operate in other sectors of society, which have deep repercussions and profound reverberations on the political sphere.

Fieldwork in social sciences, especially when dealing with informants whose lives are at stake if a controversial or highly polluting complex gets built, can be rewarding and upsetting at the same time. One of the most fascinating discoveries I stumbled upon, while doing research in Taiwan, was the variety and heterogeneity of the actors involved in green politics. While I did not always agree with their goals or strategies – even though my opinion is of no importance since I was there as a social scientist, dedicated to keeping a degree of objectivity that many of my sources found strange if not utterly frustrating – I have to admit that I grew emotionally attached to many of my informants and I feel a deep sense of regret whenever one of their fights is lost. This book represents a way to raise attention to the numerous injustices and rightful struggles on behalf of a better environment and a less polluted Taiwan.

What I hope to achieve with my work is to provide the reader with a ground-level perspective that, in my opinion, adds some touches of color and reveals more about the numerous social transformations and changes in quality-of-life issues than any conclusion drawn from statistics alone. While I cannot deny that the fieldwork had its difficulties, for example in Changhua County (彰化縣 *zhanghua xian*) where the majority of my interviewees only spoke *minnan* (閩南) dialect, which I do not understand, I have to admit that even in those moments I never felt any of them either giving up or losing patience and interest towards the crass and non-diplomatic foreigner poking her nose into very local – and often private – matters. Au contraire, locals, activists, scholars and intellectuals with whom I cooperated for all of the cases, always patiently tried their best to make me understand the behind-the-scenes, intertwined relations and complicated networks that have the potential to turn a case into a success or into a failure with life-changing consequences.

Acknowledgments

My hope is that this book will contribute to further the discussion and thinking for those involved in trying to find new practices able to integrate economic development and environmental protection, so that the beautiful island of Taiwan may once more proudly carry its nickname of "美麗島 *meilidao*."[1]

In my study of Taiwan, a prolonged and marvellous journey that lasted one year (2011–2012), I have received support and advice from many people. Paul Jobin (彭保羅), Linda Arrigo (艾琳達) and Winston Dang (陳重信) all generously shared their contacts with me as well as their precious insights as long-time connoisseurs of Taiwan's environmental arena in ways that have proven to be critical and fundamental for my research.

Many Taiwanese environmentalists, activists, scientists, scholars and officials have been a huge help and have – throughout our numerous interviews – patiently and meticulously sketched out the world of Taiwan's environmental politics for me. Among them, to name but a few, I am hugely indebted to Prof. Tang Ching-ping (湯京平), Prof. Ho Ming-sho (何明修), Prof. Tze-luen Alan Lin (林子倫), Prof. Peter Wushou Chang (張武修), Prof. Chou Chang-hung (周昌弘), Prof. Wu Kuen-yuh (吳焜裕), Prof. Chou Kuei-tien (周桂田) and, finally, Lai Fen-lan (賴芬蘭), whose friendship and shared vegetarian meals have not only enriched my understanding of Taiwan and its environmental problems, but also blessed me with a true friend. A particular thank you goes to two young women, both activists and lawyers, Lu Shih-wei (陸詩薇) and Tsai Ya-ying (蔡雅瀅), who have helped me navigate through uncharted waters, when analyzing the complicate maze of environmental regulations in Taiwan. Chen Hsiao-hsuan (陳曉萱), with her inquisitive observations and her flawless English, has helped me translate those passages and legal documents that I struggled to make full sense of on my own; to her goes my deepest gratitude.

But most of all, a special thank you goes to the numerous NGOs, concerned associations and activists I have interviewed during my fieldwork and without whom none of this would have been possible; for space reasons I cannot name them all but among the most significant ones are: Green Citizens Actions Alliance (GCAA), Taiwan Rural Front (TRF), Taiwan Environmental Protection Union (TEPU), Primordial Law Firm (PLF), Wild At Heart Legal Defence Association (WAH), Homemakers United Foundation (HUF), Society of

Wilderness (SOW) and the Green Party of Taiwan (GP). Among individual activists, my biggest thank you goes to Tsui Shu-hsin (崔愫欣), Wang Chung-ming (王鐘銘), Jenner Lin (林金保), Lai Rong-xiao (賴榮孝), Chen Man-li (陳曼麗), Robin Winkler (文魯彬), Tsai Pei-hui Frida (蔡培慧), Lin Le-hsin (林樂昕), Wu Sheng (吳晟) and Pan Han-sheng (潘翰聲), who have generously accepted to be interviewed by me several times and often involved me in many of their activities and campaigns. Finally, my huge gratitude goes to Liu Hsia-Yang (劉夏泱), a Green Party member and a Ph.D. student of ancient Greek philosophy at Taiwan National University back in 2011, who organized a weekly study group at the Green Party's headquarters in Taipei and who, ever so generously, always invited me to participate and share my opinion on most domestic matters.

This book also strives to be a thank you gift for the Taiwanese people; during the 12 months of my fieldwork I witnessed such dedication and true passion towards the environment, which has enriched my personal experiences and positive impressions of Taiwan, its citizens and their culture. Furthermore, my appreciation goes to the editorial team at Routledge, for their invaluable support and precious advice in editing and polishing the book, undoubtedly making the final version of this manuscript a much better read and to Dafydd Fell for the enthusiasm and interest he has always shown towards my research. My manuscript has benefited greatly from the constructive suggestions for revision, offered by two anonymous reviewers.

Last but not least, my biggest thank you goes to my partner, Simone; without his patience and unswerving support this book would not have been written. He has always been encouraging, solved several technical problems with my computer and has very often taken up the role of sole care-taker to our newborn son, Orlando, in order to allow me to complete this manuscript.

Financial support for the fieldwork preceding this book has come most significantly from the Swiss National Foundation (SNF). The Centre for Chinese Studies at Taiwan National Library has also contributed travel and research funds as well as a five-months research fellowship grant.

I would like to thank the *Journal of Current Chinese Affairs* for kind permission to reprint a revised version of "Development vs. Environment in Taibei," co-authored with Tu Ping-lan (涂萍蘭), as Chapter 6 in this book.

Finally, I have to note that the responsibility for any mistake and error in this book is mine only and I apologize to anyone whom I have unknowingly forgotten to thank.

Note

1 "*Meilidao*" (meaning "beautiful island") or "Formosa," is the name that Portuguese settlers gave to the island in the sixteenth century, upon discovering the beauty of its natural surroundings.

Note on Romanization

Numerous controversies surround the system of Romanization for Taiwan, with frequent debates and accusations arising in case a writer opts for either the Wade Giles or the *Hanyu Pinyin* system; adding to the confusion is the fact that many names employ a Romanization that does not tally with any officially recognized system. In this book I have decided to use the Romanization employed by the *Taipei Times* for well-known Taiwanese figures, followed by traditional Chinese characters, the first time they appear in the text. For personal names of professors, scholars, activists and the rest of my interviewees I have opted for the Romanization version of their choice followed by the traditional characters. For famous places I use the standard Romanized form in usage in Taiwan, again followed by the Chinese version. In all other cases I have used *Hanyu Pinyin* for personal names, places and the transliteration of Chinese phrases.

Abbreviations

2–28	The 1947 February 28 Incident
AEC	Atomic Energy Council
CCP	Chinese Communist Party
CSSTA	Cross-Strait Service Trade Agreement
DOH	Department of Health
DPP	Democratic Progressive Party
ECFA	Economic Cooperation Framework Agreement
EIA	Environmental Impact Assessment
EJA	Environmental Jurists Association
ENGO	Environmental Non-Governmental Organization
EPA	Environmental Protection Administration
GCAA	Green Citizens Action Alliance
GLT	Green Legal Theory
GP	Green Party
HUF	Homemakers United Foundation
KMT	Kuomintang
MENR	Ministry of Environment and Natural Resources
MOEA	Ministry of Economic Affairs
NCM	Nature Conservation Movement
NGO	Non-Governmental Organization
NP	New Party
NPO	Non-Profit Organizations
NPP-4	Nuclear Power Plant number 4
NRC	Nuclear Regulatory Commission
NT$	New Taiwan Dollar
NTU	National Taiwan University
PLF	Primordial Law Firm
SOW	Society of Wilderness
TEIA	Taiwan Environmental Information Center
TEPU	Taiwan Environmental Protection Union
TRF	Taiwan Rural Front
TSU	Taiwan Solidarity Union
WAH	Wild at Heart
WANO	World Association of Nuclear Operators

1 Introduction to environmental politics in Taiwan

Post-2008 resurgence of civil activism

Since the lifting of martial law in 1987 Taiwan has slowly transformed itself into a democratic country and its civil society has boomed; nevertheless, according to various scholars and Taiwan watchers, since 2008 there has been a partial return to semi-authoritarian methods on the part of the state (Ho, 2010a: 17)[1] even more so, according to others (e.g., Cole, 2014), after President Ma's second term started in 2012. However, rather than being pressured into silence, this unfavorable condition has prompted social movements to protect their rights with renewed vigor (Ho, 2014a: 11–12) getting them "re-started" (Hsiao and Ku, 2010). This renaissance period has its symbolical birth in the Wild-Strawberry movement[2] (野草莓運動 *ye caomei yundong*), which took place in November 2008 and has, since then, undergone two peaks coinciding with two "external" strategic events: one took place in 2011, which is coincidentally also the year of several "grassroots revolutions" that threatened to overturn numerous non-democratic regimes around the world,[3] and one in 2014.

For Taiwanese 2011 was a particularly emotional year under many aspects. Not only was this a year of celebration and national pride for the centennial anniversary of the foundation of the Republic of China,[4] but this was also the year of the shattering earthquake of magnitude 9.1 that hit Japan, which brought devastation and death onto the country and its people. Taiwan, geographically close to Japan and an earthquake-prone country itself (out of four nuclear plants, three are built in the immediate vicinity of seismic faults), developed a lively discussion on the nuclear energy issue, which gave environmentalists new vigor to fight for the complete acquittal of nuclear power (Shan, 2011a, 2011b). On April 30, 2011, several thousand people gathered in the streets of Taipei and Kaohsiung to protest against nuclear energy lobbies, in the aftermath of the devastating effects of the Fukushima nuclear catastrophe (福島核災 *fudao hezai*) in Japan (11 March, 2011). A special issue of contention centers on Taiwan's controversial nuclear power plant number 4 (henceforth NPP-4), which has currently been frozen until a referendum will settle its fate and is set in an area known for its – still active – underwater volcanoes, at a distance of about 40 km from the capital, Taipei.[5]

Moreover, 2011 was also the year in which the deal regulating the deposit of nuclear wastes on Orchid Island expired. Needless to say, even though the rent contract expired on December 31, no action was taken by Taipower[6] to remove the storage deposit, igniting further protests, due to the heightened awareness brought about by the Fukushima incident.

Apart from the nuclear energy controversy, 2011 was also a year of extreme drought – the worst in Taiwan since 2003 (*The China Post*, 2011a, 2011b) – causing the immediate implementation of a strict water-rationing scheme that brought on severe consequences for Taiwan's high tech industries and chemical plants alike.[7]

In 2011 Taiwan also experienced its biggest-ever food scare beginning in mid-May, as government inspectors testing sport and soft drinks detected dangerous levels of industrial plasticizers, which have since then been found in a great variety of other products, some of which were exported to other Asian and Southeast Asian countries, making this an international crisis with huge economic losses for Taiwan, whose products were withdrawn from supermarket shelves in mainland China, Hong Kong and the Philippines (*The Economist*, 2011).[8]

Most importantly though, 2011 was a highly significant year for Taiwan's citizens, politicians and green activists, being the year preceding the presidential elections held in January 2012 (Fell, 2011; Schubert, 2012). This means that

Figure 1.1 A dry river in Taitung, May 30, 2011 (source: Simona Grano).

every move, every decision and every public display of outrage (or consent) towards both the environment and potentially ecologically damaging projects (from nuclear power plants to petrochemical complexes and sporting facilities) were mostly driven by ideological and political motivations and were exploited and manipulated, with the not-so-hidden agenda of attracting votes, by all the major parties present on the political scene.

It is thus sufficiently safe to conclude that 2011 represents a watershed year in Taiwanese politics, during which the status of environmental issues was elevated and brought to the fore of the political agenda. In fact, while until a few years ago environmental concerns featured prominently in local elections with issues such as national identity and cross-strait relations dominating national level politics (Fell, 2013), a reversed trend has been established since 2011, which is clearly visible from the increased attention paid to nuclear energy issues after Fukushima, as well as from the growing number of TV shows (e.g., "Our Island," 我們的島 *women de dao*) and popular TV series ("Heroes of the Country," 國民英雄 *guomin yinxiong*) featuring environmental pollution, developmental projects and developers' corruption among their *leitmotivs*. This does not mean that environmental problems were the most relevant topic discussed in the pre-election campaign period (China's growing influence over Taiwan was a far more relevant topic); however, it does indicate that such concerns have become more significant than they ever were in the past. Furthermore, in a period in which social unrest was brewing already, Taiwanese activists united their efforts under the common banner of requesting more transparency and accountability on part of the government, thus increasing the status and visibility credited to ecological concerns. Political parties, aware of the increasing importance of the powerful pro-green rhetoric, exploit such issues to attract votes, often without a real commitment. This signals a change from the past, when political campaigns were almost exclusively focused on unification with the PRC and economic development (Fell, 2013). However, as we shall see with the analysis of the four case studies, environmental activists have also learned how to make use of this powerful rhetoric and of the specific, strategic time in which events take place, in order to achieve their goals.

The Fukushima disaster, in all of its drama, has contributed to revive long-dormant anti-nuclear feelings, causing a major change at the political level, with a traditionally pro-business party such as the Kuomintang – KMT – (中國國民黨 *zhongguo guomindang*) reaching the decision to halt construction of the nearly completed NPP-4. Whether this could indicate a shift in the KMT's perception of what concerns Taiwanese citizens or whether such issues have become so powerful that no political party can ignore them, lest they risk the loss of future electoral battles, remains to be seen. What is certain though, is that the stronger emphasis on green issues during the pre-election campaign period indicates a novel trend, favorable to environmental protection, which has been established since this watershed year and has been growing ever since. 2011 was definitely a remarkable year not only for the anti-nuclear movement, but also for the environmental movement at large with the successful and formidable opposition against the Kuokuang naphtha cracker, a case study, which will be expounded in Chapter 4 of this book.

Finally one last observation in regards to the future of environmental protection in Taiwan; currently "nature conservation" is a prominent concern of officials, common citizens and activists, as can be observed in the numerous leaflets heartily portraying the island's natural beauty and wild parks, as well as in the heightened awareness towards "ecological problems" such as water conservancy and recycling activities. While for a long time such issues represented nothing more than careless topics for superficial, over-dinner conversations, recently Taiwan has experienced an upgrade of efforts and commitment towards the protection of its fragile ecosystem. To quote Ho Ming-sho paraphrasing Ulrich Beck's essay on the "world risk society" (Beck, 1986) "Perhaps, the worldwide sympathy for Japanese people and resurgence of anti-nuclear movement following the wake of the Fukushima disaster in March 2011 constitutes an instance of cosmopolitan consciousness in the era of world risk society" (Ho, 2012: 3).

Building on Ho Ming-sho's own categorization of social movements in Taiwan from the 1980s until today (Ho, 2010a: 16), social movements underwent a phase of resurgence (2008–2010) after a decade of "institutionalization" coinciding with Chen Shui-bian's administration, in which no remarkable protests erupted. I contend that this resurgence phase has not yet exhausted its force but that, actually, it has been growing stronger since the year 2011.

In fact, Taiwan's political environment between 2010 and 2012 was extremely volatile with the country undergoing two mayoral and one presidential election.

The second Ma Ying-jeou (馬英九) administration finds itself in the uneasy situation of having to balance waning popular support with the implementation of unpopular trade agreements (Chow, 2011: 185–186); protests are directed at a plethora of issues of social discontent: from those who claim that the Economic Cooperation Framework Agreement,[9] enacted in 2010, is more favorable to China than Taiwan and serves to advance China's unification agenda, to the numerous anti-nuclear rallies, which have become a frequent feature of Taiwanese street life, to accusations against President Ma of having exploited the Wang Jin-pyng (王金平) affair to further his personal interests in trying to oust a fellow KMT member and political rival. However things may be, these past few years have been characterized by a vigorous civic activism with several large-scale protests and demonstrations in 2013 alone (Wang, 2013a). Protests were not only environmental of course, ranging from indignation over the unclear circumstances surrounding the death of a young army corporal who was being punished for misconduct (Hung, 2014: 54–77), to the long-going anti-media monopoly dispute (反媒體壟斷 *fan meiti longduan*) in 2012–2013 (Harrison, 2012; Rawnsley, 2014b). Younger generations, who are the major force behind the social unrest, have been taking to the streets to defend issues such as human rights, social justice and environmental protection, while openly battling governmental corruption and abuse. The Sunflower Student Movement (太陽花學運 *taiyanghua xueyun*),[10] which occupied the legislative floor for several weeks to protest against the Cross-Strait Service Trade Agreement (CSSTA) in March 2014, is the most prominent example of such unrest and epitomizes the culmination of the several protests that have taken place since 2008 (Cole, 2014). All of this represents an unprecedented re-awakening of civil society, in the history of democratic Taiwan, which has, at the same time, contributed to re-launch several other issues, such as environmental concerns and anti-nuclear sentiments or public awareness in regards to the government's willingness to promote economic agreements with China, that are considered by some as tools to advance Beijing's reunification goals (Yuen, 2014: 74). Amidst the growing uneasiness and discontent towards Ma Ying-jeou's administration, barely two weeks after the Sunflower occupation of the Legislative Yuan, former head of the DPP, Lin Yi-hsiung (林義雄), launched a long-planned hunger strike in favor of the abolition of NPP-4, which brought several thousand people to the streets, who set up blockades on public arteries and organized a series of anti-nuclear rallies. Against the backdrop of rising social unrest, this book tries to sketch the most salient points of the environmental policymaking arena in contemporary Taiwan, positing particular attention at the newer cohorts of environmental activists that have emerged in the past few years, and are often characterized by a more radical positioning (by no means violent) while at the same time displaying a far more professional approach and a deeper understanding of the available mechanisms to fight developers and polluting facilities. These can range from a skilled employment of the legal system to a deeper effectiveness in spreading their ideas through online platforms and forums. Paraphrasing Michael Cole (2014), smaller rallies organized by this new generation of activists, though not as flashy as larger protests usually portrayed by the media in their reportages, have been much more successful at generating responses on the part of the government and consequently pose a greater concern for the authorities who have tried to deter them or discredit them. One of the elements of success of these newer groups is their ability to unite under a common banner, which transcends older divisions that often alienated one group from the other in the past, such as ethnic, generational and political divisions, giving shape to a truly unified opposition movement. This movement has shown its full strength by urging the general public to remember that the KMT's governance over the past year has been characterized by infringements on human rights, forced evacuations, demolition of homes and the forceful seizure of land, among others. Whether these allegations are founded and represent the real motivations for the widespread discontent or whether the public is unhappy about the country's situation of economic stagnation, the DPP, as well as a few independent candidates (e.g., Taipei City's mayor-elect Ko Wen-je 柯文哲) have benefited from the disappointment and anger towards the KMT, which suffered a massive setback in the November 29, 2014 municipal elections,[11] when the public revoked its mandate in several key strongholds and gave it to the DPP instead. These elections were a victory for smaller parties like the Green Party, who managed for the first time since its establishment in 1996 to win two seats out of nine candidates it had presented and for the newly established Tree Party – a GP splinter organization founded in 2014 by former GP secretary-general Pan Han-sheng (潘翰聲), which triumphed in the mayoral election for Nantou County's Chichi Township (集集), marking a successful start for the party. Such gains seem to indicate a growing awareness and support for parties that prioritize environmental awareness and grassroots social activism.

Reasons for writing a book on environmental governance in Taiwan

> Scholars who analyse political power can be divided into two groups, each with a different perspective. One group emphasizes social conflict; the other, social cooperation. Depending on the emphasis, what the two see as the nature of political power is quite different.
>
> (Fei Xiaotong, 1992: 108)

The contemporary world faces a wide selection of difficult social and environmental problems that often transcend established social and territorial boundaries. Therefore, modern states try to reorganize involved institutions in the hope of finding ideal solutions for effective governance. This has led to the emergence of a host of new actors, of new forms of state-society relations, of hybrid public institutions and networks of multi-sector partnerships, while questions of accountability and legitimacy remain unresolved. These developments necessitate a fresh look at the question of governance.

The idea of writing a book on the topic of environmental governance in Taiwan began after my reading of *Taiwan's Environmental Struggle: Toward a Green Silicon Island* a few years ago. While this impressive work addresses several significant ecological disasters and issues, facing Taiwan during the past 50 years, the analysis terminates at the beginning of Chen Shui-bian's second presidential term. However, a great deal has changed since then and several promises for a better environment have been betrayed. With Ma Ying-jeou's administration well into its second term, it is a timely topic to evaluate the recent changes that have taken place in environmental politics in Taiwan. Furthermore, an aspect that is hinted at but not fully developed in that pioneering research is the strong connection between politics and ecological issues in Taiwan, which will constitute the common thread linking the various case studies in this book. This often unnoticed and yet very strong link is present in all of the developmental projects analyzed and in many other environmental struggles and fights, which due to space constraints I was not able to include and has varied and far-reaching consequences for environmental protection; in fact, without strong governmental will, nothing can be achieved. Besides, the recent developments which have taken place in the civil society galaxy in Taiwan, with the partial linking of the anti-nuclear cause with the Sunflower Student Movement's agenda (and fear of the PRC's rising influence over Taiwan lurking in the background), necessitate a fresh look at the younger (greener) wing of Taiwan's environmentalism and at the potential political fallouts of such a vigorous civil activism, not just for Taiwan itself and the KMT legislature, but for two international powers such as Washington and Beijing.

There are several reasons why a closer look at Taiwan's environmental governance is warranted. The country underwent a challenging transition from industrial to post-industrial society, in which environmental quality slowly became a top priority, after decades of reckless pursuit of profit. It earned the "economic miracle" name in the 1980s, experienced a "silent revolution" in the direction of democratization in the 1990s and cultivated a prosperous civil society in the new millennium, which fights heartily for many causes, among them the environmental one.

The painful lessons Taiwan has learned throughout its journey should be of interest to other developing countries, illustrating how these positive transformations have managed to bring about a more ecologically friendly mode of economic development. Second, Taiwan possesses the most vibrant civil society and environmental movement in the economically dynamic region of East Asia. While South Korea's civil society is at times confrontational, environmentalists have so far failed in their challenge to the state's nuclear and petrochemical plans (Lee, 1999: 117; Hong, 2011), which has seen the Taiwanese environmental movement scoring major victories; topics addressed, respectively, in Chapters 3 and 4. The case of environmental politics in Taiwan is therefore of great interest for research on environmental politics at large, in comparison with other countries, China in particular. Its relevance as a model for China to solve its environmental issues should not be underestimated; both states share authoritarian pasts. As Taiwan first embarked on the path of democratization, its environmental movement played a crucial role. China is just about starting to recognize and tackle the costs of its economic success through a mixture of popular protests, public activism and governmental programs. Taiwan's experience, as amply shown in Chapter 2 of this book, is therefore of the utmost importance seen in this light. Furthermore, the Sunflower Movement Occupation has altered the awareness that Taiwanese people have vis-à-vis the "China Outside"[12] (Lowther, 2014), so that even those previously "neutral" organizations and movements (e.g., the environmental ones) are now conscious of the international dimensions of their struggle and its meaning for the future of the island.

In fact, a rarely talked about fallout of the CSSTA is its impact on Taiwan's environment. This stems from the country's environmental regulations system, which contracts out environmental impact assessments to external environmental-consulting companies. The fear among environmentalists is that if the CSSTA is passed the service industry related to environmental issues could also be opened up to Chinese investors, about whose professionalism they have serious doubts. According to Echo Lin, secretary-general of the *Environmental Jurists Association* (環境法律人協會 *huanjing falüren xiehui*)[13] – henceforth EJA – as subcontractors of the Taiwanese government such companies would have privileged access to confidential information regarding Taiwan's geological situation, which could pose a threat to the country's national security (Echo Lin, email communication, October 24, 2014).

In recent years, both in scholarly as well as in political debates around the world, attention has shifted increasingly towards local governance, which rather than focusing solely on the local government, indicates that decision-making processes involve multi-agency and multi-level collaborations, networks and partnerships. While governments remain the leading public stakeholders, a greater emphasis is nevertheless placed on participatory democracy.

While Taiwan has long been experimenting with the creation of legal channels for citizens' participation during planning evaluations, such participatory channels are not devoid of problems and activists and social groups pinpoint several of their flaws. Thus, this book will posit particular attention to the legal channels and rules-based participatory approaches employed by activists – as well as by governmental administrators at various levels – to either oppose or promote projects.

Taiwan is often heralded as a brilliant example of a country that has enjoyed a smooth transition to democratization; however, an analysis of its green politics will show how democratization, by empowering agents that previously lacked power, has created both opportunities and obstacles for those groups whose interests lie with environmental protection (Tang and Tang, 1999: 351). In Taiwan, all political parties prioritize economic growth over environmental protection. The two main parties present on the political scene use the environment as a tool to attack each other, the Democratic Progressive Party – DPP – (民主进步黨 *minzhu jinbudang*) claiming that the KMT's pro-growth stance is irresponsible, and the KMT maintaining that the DPP is anti-business. However, Taiwan is also a place where regulators and city officials have attempted to redress certain imbalances in the ecological environment by enacting a series of programs aimed at the creation of a green, sustainable economy and a recycling-oriented society (Her, 2010: 4–11). While environmentalists gained access to policy decision-making after the democratization process started in 1987, pro-development groups have also increased their power since then, and government policies have consequently shifted back to a more pro-business position. As argued by Ho "this paradox resulted from the weakened state capacity, which simultaneously empowered environmentalists and could not resist business's lobbying" (Ho, 2005a: 339).

The numerous "stakeholders" participating in the environmental governance process have been the focus of public policy research but have seldom been studied in terms of their political power and their ability to influence the implementation of green policies. Furthermore this book will focus in particular on a new type of stakeholder: the lawyer/activist, who has emerged as one of the leading figures of Taiwan's environmental struggle in the past few years.

The main argument put forward in this book holds that although Taiwan needs civil and official collaboration on environmental protection, most of the time the implementation of an efficient system of environmental regulations is hindered by many agents and stakeholders, involved in the process of environmental governance (state, economic actors and institutions, civil society, media, recently re-awakened aboriginal consciousness, etc.), whose interests greatly diverge.

Although environmental policies are, for the most part, mandated from the top, at the local level their implementation is altered by the interaction of these numerous agents. Consequently, there is no uniform, homogeneous authority for environmental governance but rather a variety of official and less official agents of authorities whose interests and powers overlap and compete. Environmental governance, being a multi-layered process involving many actors and stakeholders can be, in turns, dominated by powerful business actors prioritizing economic growth over ecological sustainability or it can be more inclusive, thus representing marginalized human and natural constituencies (Whitman, 2008: 164). This multi-scalar dimension of governance will receive special attention in the four case studies analyzed, in order to determine which agents are more successful at influencing environmental regulations in loco and through which means.

More specifically, environmental governance is a complex process involving various environmental agencies, industrial bureaus, local governments and many other governmental units at both the central and local level. This process is often characterized by a conflicting relationship between lower and upper ranks of government and between pro-business factions and environmentalists. One way to gain the upper hand in this complicated and fragmented governance process, is to build up consensus among the general public. Consensus-building is in fact of crucial importance both for lower levels of government trying to push forward a developmental project boosting local revenues, as well as for activists trying to stop it, with public opinion often turning into the ultimate deciding factor. How each actor tries to exploit and win over the public – through which methods and strategies – will also be in the focus of the present study.

As previously mentioned, in the past decade, with a strong increase in the past five years, Taiwan has experienced an upsurge in the availability of rules-based participatory approaches through which citizens play an active role in policy mechanisms, while pressuring local governments and developers into compliance. This positive turn represents a barely researched aspect of Taiwanese green governance, which will increase our understanding of how such mechanisms work and, hopefully, contribute to achieve a more effective management of both natural resources as well as man-made facilities. Furthermore, while the main thesis of this book, that environmental governance includes multiple actors and stakeholders, is per se not entirely novel, all the developmental projects analyzed represent special cases of the above-mentioned new category of stakeholders, which has emerged in Taiwan in the past few years, of lawyers/activists, who lead the opposition movement towards polluting facilities in a professional way, through the employment of the legal system, corroborated by a profound knowledge of the most poignant pieces of legislation. Since a commonly employed state's refrain is calling activists "anti-democratic" when they engage in direct action "outside the legal structure to challenge authority" (Rawnsley, 2014a) the employment of those very same legal channels set up by the state, on part of these lawyers/activists, dispels any such critique. Furthermore, the younger wing of Taiwan's environmentalism, with its new unconventional methods (e.g., rock concerts and cyber activism) and more radical stance, which has emerged as a significant key player in all the most salient protests of the past few years, explains the sudden interest newer generations have shown towards ecological concerns, due to the trendy appeal of cultural activities such as concerts, documentaries and poetry readings, and prompts renewed attention towards the study of Taiwan's environmental governance structure. In fact, the prominent role of the country's artistic community in all the most relevant environmental protests of the past few years represents a novel and unexpected turn of events for a country in which, for a long time, the artistic community was either apolitical, socially conservative or cautious in its critique (a remnant of the KMT's total control over Taiwan's cultural production for several decades).

As we shall see in Chapter 4, in Taiwan environmental risks and harms affect certain geographic areas and ethnic groups more than others. The environmental movement tries to force public and private sectors to address this disproportionate burden of risk and exposure to pollution in minority and/or low-income communities (e.g., Changhua County), and strives for those communities to be engaged in the decision-making processes through channeling their fight inside the legal system, rather than simply being forced to accept top-down regulations. Such methods of contention are often clustered under the umbrella etiquette of "Green Legal Theory" (GLT), which in contrast to classical environmental-law approaches emphasizes the limits of relying solely on state-issued and top-down regulations; these, more often than not, turn out to be socially constraining (M'Gonigle and Ramsay, 2004). The GLT approach tries to conceptualize how legal and political institutions could offer more than a plethora of laws and regulations aimed at protecting nature in favor of fostering a new sense of awareness on what is missing in order to create self-sustaining (and sustainable) social, economic and political institutions, which recognize how traditional actors (modern liberal states and societies, and their environmental regulations) are often tied to unsustainable practices. As mentioned, such a situation has facilitated the appearance and gradual establishment of many legal experts involved in environmental struggles, whose commitment and engagement goes way further than that of a mere professional offering its services, being they themselves activists on behalf of ecological protection. While such figures are not entirely new in countries where an environmental movement is more developed, such as the United States and Germany, in Taiwan they represent an absolute change with the past in the direction of a more sophisticated battlefield in which actors are experts with a profound knowledge of the available legal tools. Their engagement has reached new heights of professionalism and effectiveness with the establishment of EJA in 2010, which offers legal services to those in need, in cases in which natural areas deemed to be in the public interest, are threatened by a polluting facility and hence an environmental damage might arise. According to J. Michael Cole, a Taipei-based journalist, one-fifth of Taiwanese lawyers are active in offering pro bono legal advice and giving out practical information on how to act in difficult situations (e.g., in case of arrest during a protest), following the big increase in civil disobedience that the country has been experiencing in the last 5 to 6 years (Cole, speech at SOAS, June 16, 2014).

Research questions and methodology

Social sciences can help us understand the complex interrelation taking place between society, politics, economy and the environment (Yearley, 1991: 184). While natural disasters and problems require scientific knowledge of natural sciences, this book will show how, through the study of relevant case studies, the greatest damage to the natural world is often brought about by humans; environmental problems are thus nothing else than a reflection of social problems, short-sighted development and corruption and can thus be studied (as well as partly solved) through the lenses of social and political sciences.

This study will demonstrate that even though environmental rules and laws are written at the central level, in reality, a multitude of other issues and agents can alter their enforcement at the local level. Much regional activity, including regional environmental management, is in fact the result of resources, opportunities and interactions taking place at other levels. Furthermore, local governments and other actors strategically manage and employ multi-scalar relations of environmental governance in order to promote or impede specific public policies and social movements' (green) initiatives. In other words, independent but connected political decision-making centers, acting within their respective areas of competence, can influence and alter the correct functioning of an environmental governance system. These issues will be analyzed in a multi-disciplinary perspective that looks at central-peripheral relationships, local practices and institutional configurations as well as the function of social norms and cultural value systems.

First, a key question this research will try to answer is how does environmental governance work at the local level and which actors are more influential in pushing forward their agenda: traditional power-holders such as business elites or new stakeholders such as activists and concerned residents? A major question for the research of the case studies analyzed has been whether social movements' and other actors' environmental campaigning strategies have been affected by domestic political issues. This is particularly interesting in the case of industrial pollution issues (nuclear and petrochemical) and the role they occupied in the pre-electoral campaigning year of 2011. Second, the book aims at challenging the commonly known assumption that "environmental issues" are only remotely connected to politics by showing how the link between "green issues," social movements and the course of politics has become stronger in the past decade in Taiwan, albeit remaining in many instances still a mere smoke-screen to attract voters. In fact, as we shall see, political parties, aware of the increasing importance of the powerful pro-green rhetoric in the past few years, exploit such issues to attract votes during pre-electoral periods, albeit – often – with no real commitment.

Field research for this book was conducted in Taipei City, Taipei New City, more specifically Tamsui municipality and Gongliao District, Changhua County and Pingtung County (屏東), namely Kenting (墾丁), from February 2011 to January 2012. Follow-up interviews via email and updates on the most recent events were carried out throughout 2013–2014. The majority of the civil society organizations interviewed are grassroots NGOs except for one, which is sponsored by the local Tamsui (淡水) government, namely the Tamsui Heritage Foundation. For space reasons – and much to my regret – I cannot name all of the organizations interviewed here that have made invaluable contributions to this study; among the main grassroots organizations I talked to are: *Green Citizens Action Alliance* (綠色公民行動聯盟 *lüse gongmin xingdong lianmeng*), the *Taiwan Society of Wilderness* (荒野保護協會 *huangye baohu xiehui*),[14] *Taiwan Environmental Protection Union* (台灣環境保護聯盟 *Taiwan huanjing baohu lianmeng*),[15] *Taiwan Rural Front* (臺灣農村陣線 *Taiwan nongcun zhenxiang*),[16] the *Mothers' Alliance for Monitoring Nuclear Power Plants* (媽媽監督核電廠聯盟 *mama jiandu hedianchang lianmeng*),[17] *No Nuke Cultural Activism Group* (諾努客文化行動團隊 *nuo nuke wenhua xindong tuandui*),[18] *Organization of Urban Re-s – OURs –* (專業者都市改革組織, *zhuanyezhe dushi gaige zuzhi*),[19] the *Songshan Tobacco Factory Park Union* (松山公園催生聯盟 *songshan gongyuan cuisheng lianmeng*), the *Taiwan Environment Information Center* (環境資訊中心 *huanjing zixun zhongxin*), *Homemakers United Foundation* (主婦聯盟環境保護基金會 *zhufu lianmeng huanjing baohu jijinhui*).[20] Several members of Taiwan's Green Party – GP – (綠黨 *lü dang*) have proved an invaluable source of information on the behind-the-scenes of green politics and have provided links and contacts with numerous concerned legal associations, working on behalf of environmental protection such as *Wild at Heart Legal Defense Association* (臺灣蠻野心足生態協會 *Taiwan manye xinzu shengtai xiehui*),[21] the *Primordial Law Firm* (元貞聯合法律事務所 *Yuanzhen lianhe falü shiwusuo*),[22] the *Lin and Shi Legal Office* (永信法律事務所 *Yongxin Falü Shiwusuo*) and of course the newly established legal association devoted to environmental protection the *Environmental Jurists Association*, all involved in several of the cases analyzed in the book.

The methodology employed was inspired by that used in similar studies in other parts of the world, and it hinges predominantly on extensive usage of interviews (informal, unstructured interviewing and semi-structured interviewing) with concerned activists, local residents and related local institutions, as well as on a review of relevant literature. The extensive fieldwork, carried out in one prolonged stay, included in-depth interviews with activists from more than 20 environmental organizations and institutions and dozens of public-service employees, Environmental Non-Governmental Organizations' (ENGOs) representatives, lawyers and health hazards experts, attendance at and participation in more than 30 lectures, workshops and seminars organized by civil society organizations, universities, public interests groups, as well as in public hearings for environmental impact assessments.

This book draws on qualitative empirical data from four protests, which took place between 2010 and today, three of which are ongoing, yet-to-be-completed projects as of 2015. I also participated in on-site visits and informal social gatherings together with activists involved in the different campaigns. Documentary sources in the form of petition letters and reports written by protesters were consulted throughout the research. Moreover, I carried out multiple interviews with several academics and scholars with in-depth knowledge of the various cases. Participant observation methods were employed in two of the developmental projects analyzed, the Taipei Dome (Chapter 5) and the Tamsui North Shore Road Project (Chapter 6), with myself actively participating in all of the activities (from the organizational start-up phase to the actual street protests) organized by several of the main social groups devoted to fight the two projects, aided by the Green Party of Taiwan.

I am aware that the depth of my involvement might lead to a certain bias in the study, and that it will be difficult to give expression to a critical perspective regarding the specific qualities of social movements' discourse; everybody is green or wants to be green right now, whatever that means – though business is going on as usual, green or not. Nevertheless, I have to point out that it is precisely my active participation that has helped me to delve into these protests and discover facts that might have been neglected or forgotten if I had restricted my role to a more detached and distant observation. Furthermore, by including several stakeholders from all factions and backgrounds, I believe I have been able to maintain a relatively objective perspective.

For the first case study analyzed in the book, dealing with nuclear energy and the construction of controversial NPP-4 (Chapter 3), I analyzed a mixed body of data including case studies by medical and health professionals (in the case of nuclear wastes dumped on Orchid Island), surveys by governmental agencies and researchers, investigative reports and news articles in printed and online media and ethnographic first-hand evidence obtained via interviews and participant observation during the fieldwork.

For the second case study, constituting Chapter 4 in the book, regarding a petrochemical project in Changhua, I interviewed numerous grassroots NGOs, two legal firms, the Green Party of Taiwan, a toxicology expert from National Taiwan University (NTU), an Academia Sinica scholar of plant ecology, several local residents of Changhua County – whose farmers and fishermen would have lost everything had the naphtha cracker been built – and a few writers and prominent film directors involved in the protection of the wetland on which the petrochemical plant should have been built.

In Chapter 5 I analyze the construction of a sporting facility threatening an historical relic and its lush green park in downtown Taipei, focusing on the political issues and the behind-the-scenes of a project that is being built in one district where the Green Party was running alone for the 2012 elections, Xinyi District (信義區).

Finally, for the fourth case study, which constitutes Chapter 6 in the book, field research was conducted in Taipei City and Tamsui District from February to June 2011. Among the experts interviewed I have talked to several governmental officials of both Taipei City and Taipei County's involved bureaus, with local residents, with Green Party candidates running for the legislative elections of 2012 and with a scholarly expert of transportation processes. Follow-up interviews as well as participant observation were carried out in the fall of 2011.[23]

Because interviews were developed through snowball sampling, the data are not meant to be statistically representative of the environmental movement, although interviewees do come from a wide range of urban and rural areas.

All interviews were conducted in Chinese by myself.

A few words have to be spent in explaining the criteria behind the choice of the four case studies, among the numerous environmental protests that have taken place in Taiwan in the past few years.

First of all, I was looking for cases sharing similar traits in involving a multitude of agents and stakeholders as well as different levels of governments, all with conflicting and/or colluding interests: some trying to protect the environment, others trying to push forward with their developmental plans and others yet simply trying to implement the law. The significance of the two cases focusing on the nuclear and petrochemical projects is easy to grasp, being that nuclear energy and the petrochemical industry are related to industrial pollution, arguably the classical source of environmental problems; furthermore both of these cases represent landmark events that indicate critical transformation of Taiwan's societal response to environmental concerns. The other two cases sampled focus, respectively, on urban renewal/commercial development and on transportation infrastructure construction. I have deemed it necessary to choose four very different typology cases in which the environmental movement has been actively involved, through the employment of rules and regulations, so as to offer a picture as comprehensive as possible of the evolution of environmental governance in a regime under democratic transition, such as Taiwan in the age of post-industrialization.

For instance, the anti-nuclear cause has been an ideology-laden case that has pitched pro-business elites against environmental movements (sometimes with deep divisions present among environmental activists); the Kuokuang case represents a "NIMBY turning into conservation" case while the other two cases bring to the fore newer concerns such as "tree protection," "urban heritage conservation" and "ecological safeguarding," which are samples of a broader category of environmental problems present in Taiwan that are critical to advance our understanding of the country's social renovation, in an era in which environmental issues are gaining a more prominent place.

Last but not least I have deliberately selected cases in which activists interact with the legal system by making use of the available legislative channels for public participation in framing their opposition to the various developmental projects and by emphasizing illegal activities and violations of the *Environmental Impact Assessment Act* (環境影響評估法 *huanjing yingxiang pinggu fa*), henceforth EIA, on the part of the local government and the developers, aided by professional legal experts. This holds true for all cases analyzed in the book in Chapters 4, 5 and 6, except for the anti-nuclear opposition to the island's fourth nuclear power facility. However, considering the significance of the anti-nuclear controversy for Taiwan's social transformation and the high visibility of anti-nuclear protests since 2011, it would have been impossible not to include this case in my analysis of Taiwan's current environmental struggle. Furthermore, while not directly trying to use the law to challenge the construction of NPP-4, which would be pointless given that construction of the facility is by now completed, lawyers and legal teams are nevertheless involved in this case as well, assisting green groups and the DPP in their proposal to amend the country's *Referendum Act*, as will be explained at the end of Chapter 3.

Synopsis of the book

Following this introductory chapter, the book consists of five more main chapters. Chapter 2 constitutes the background part, which attempts to offer a general overview of environmental politics and issues under the KMT as well as of the main changes that took place after the DPP gained the presidency in the year 2000. The context chapter will highlight the changes (or lack thereof) in Taiwan's environmental governance under different political parties and it will also provide the reader with some crucial information in regards to different parties' stance vis-à-vis "ecological concerns" and "economic development." Furthermore, this chapter will highlight the strong link present between politics and green issues in Taiwan, while challenging the commonly used stereotype that environmental politics are only a secondary aspect of the policymaking process. In the second and final part, the analysis will shift to the origins of the environmental movement in Taiwan, its political orientation and highly politicized nature that, in the early days, took the form of dissent towards the one-party-rule hegemony that had damaged Taiwan's ecosystem (Arrigo and Puleston, 2006: 167–168). The analysis will then proceed to the main actors and stakeholders in Taiwan's environmental arena; in particular, it will focus on social movements' early struggle towards nuclear energy and island-wide industrial pollution, brought about by the reckless exploitation of a short-sighted government, which for several decades considered Taiwan as a mere temporary retreat.

As mentioned, the background thesis of the book focuses on the strong but often forgotten relationship linking green issues with wider political and social questions. To emphasize this, Chapter 3 begins with one of the contentious episodes that best exemplify the persistence of a pro-development bias on part of the two main parties present on the political scene, the already mentioned NPP-4. The analysis retraces the controversy surrounding its lengthy construction (over 20 years) and the numerous scandals that have marred this project. In fact, the fourth nuclear power facility best epitomizes the link between politics and environmental concerns and has come to symbolize the struggle against probusiness elites on the part of pro-democracy activists, since its early planning phase at the end of the 1980s.

The chapter begins by offering a brief historical overview on nuclear energy in Taiwan. It then proceeds to enumerate the numerous problems and rumors that have surrounded NPP-4 for more than two decades. Afterwards the analysis moves on to the perception common citizens have towards nuclear energy, after the Fukushima incident rekindled anti-nuclear feelings inside Taiwan. Special attention will be posited to the exploitation of nuclear issues by politicians during their pre-electoral campaigns in 2011. Finally, the last part of the chapter will briefly illustrate the motives for the KMT government to propose a referendum to decide the fate of NPP-4 and will analyze the most recent changes in regards to the temporary halting of the construction of the island's fourth nuclear facility, in the aftermath of the Sunflower Student Movement and Lin Yi-hsiung's hunger strike.

Chapter 4 discusses the second case study, dealing with the construction of Taiwan's eighth naphtha cracker, namely the *Kuokuang Petrochemical Technology Co.* (國光石化科技 *Guoguang Shihua Keji*) in Changhua County, hereafter KPT. The chapter retraces the birth of the petrochemical industry in Taiwan and its significance for the country's post-war economic development. It then centers on the proposal on the part of China Petroleum Corporation, Taiwan (台灣中油 *Taiwan zhongyou*), henceforth CPC, to build a massive petrochemical complex on one of the country's last unspoiled wetlands, threatening its fragile ecosystem and a thriving fishing industry, in order to expand Taiwan's oil-refining capacity. The second part of the chapter describes the different stakeholders and agents either supporting or opposing the project and their role in the controversy. Through an analysis of their strategies and methods, I will try to ascertain who were the most successful players in this case. A small section of this chapter is dedicated to the exploration of the cultural dimensions of environmentalism, in the attempt to discern how Western-imported ideas are reprocessed and hybridized by protesters with local traditions, through the employment of creative methods involving the artistic community's celebration of Taiwan's native maritime culture and precious natural surroundings.

Chapter 5 analyzes a sporting facility, namely the *Taipei Dome* Project, which is currently under construction in downtown Taipei and should be finished by the end of 2015. This case, in contrast with the first two, takes place in a urban setting and has by far a smaller environmental impact than the one forecasted for NPP-4 and KPT; however, it embodies a perfect example of how environmental governance is becoming a multi-layered process, where, more than actual laws and regulations, the deciding factor for a facility to obtain a green light seems to be the outcome of political ties and economic interests. The Taipei Dome case is also particularly apt to illustrate the link between politics and the environment. In fact, the complex is being built in one of the districts where a Green Party candidate was running for the 2012 legislators' elections with outward support from the DPP. Thus, the stopping of the Dome became a symbol for the Green Party, spearheaded as an example of what the GP's platform would turn out to be in the case of an electoral success. Furthermore, the arrest of Farglory Corporation's chairman, in June 2014, will be expounded in order to understand its implications for the *Taipei Dome* Project.

Chapter 6 centers on the construction of the *Tamsui North Shore Road* Project; a surface road linking Taipei City (台北市 *Taibei shi*) with Tamsui municipality or New Taipei City (新北市), illustrating the evolution of a protracted conflict involving road politics in northern Taiwan. Slightly different than the other cases, this project offers us a glimpse into local politics and into how two different municipalities with conflicting views and interests with regards to this road attempt to overcome each other. The final part of the chapter will focus on the most recent developments, which saw environmentalists scoring a major victory in having the verdict of the first EIA upturned and declared invalid by the Taipei High Court.

As mentioned, in the KPT, Taipei Dome and Tamsui Road cases, a special focus will be on the legal aspects and regulations and available participatory channels that green activists have sought to employ to reach their goals and fight these construction projects.

Finally, Chapter 7 reaches some conclusions in regards to the main issues addressed in the book and tries to answer those research questions raised in the introduction; furthermore the conclusive part will attempt to shed some light on the specific characteristics that have the potential to turn a case into a success or into a failure for the green movement. The interplay of various stakeholders, social organizations and agencies at the local and national level, the way they respond to each other and to transnational developments, are important aspects that will be analyzed throughout this study in the hope of clarifying the role of the different agents and the delicate web of entangled powers and relations involved in the environmental governance field. The analysis in the book proceeds along two trajectories: a temporal one – all of the cases have either taken place or have been experiencing a revival since 2011 – as well as a spatial one, with protesters increasingly at ease and aware of the legal rights and rules-based participatory approaches at their disposal.

Research on governance as applied to Taiwan

Comparing different analytical frameworks

The first part of this overview provides a summary of the existing literature on governance in general and more specifically on environmental governance in Taiwan. I will discuss the English and Chinese language literatures together as Taiwanese and foreign scholars, despite a different perspective on things, all read each other's works and utilize common theories, when analyzing the concept of governance.

After reviewing the most important and informative approaches to the concept of governance, the second part introduces concepts and theoretical issues that inform my own study of Taiwanese environmental governance and the analytical framework for this book. The last section deals with the specific characteristics of Taiwan's environmental governance, by focusing on its bureaucratic structure and governments' capacity in the field of environmental protection.

Fragmented authoritarianism and heterarchy

The most comprehensive and still up-to-date framework through which to study politics in China is the "fragmented authoritarianism" framework, put forward by

Lieberthal, Oksenberg and others in the late 1980s to early 1990s. This system, although created for the study of China, can just as well be applied to Taiwan's political situation when the liberalization process, introduced in the 1980s and 1990s with an emphasis on decentralization, transformed the traditional system and rendered it more "fragmented." According to this theory, policies made at the central level become increasingly "soft" and susceptible to the parochial goals of multiple agencies and actors charged with the task of enforcing that specific policy at the local level. The final policy outcome will result from the incorporation of the multiple and varied interests of these agencies in the actual policy itself (Mertha, 2008: 5). The end result is that the real policy will often turn out to be quite dissimilar from the original desires of the policymakers at the central level; this phenomenon applies perfectly to the varied and complicated reality of Taiwanese green politics. As we shall see in Chapters 4, 5 and 6 this is particularly true for what concerns the EIA act, which has been drafted in a professional way in the books, but de facto is often implemented in arbitrary ways, with a variety of parochial interests and business incentives influencing its application.

For this study, concepts like "state" and "local government" are not to be taken as unitary wholes situated at the top or bottom of the power scale. In all of the case studies chosen, the anthropological concept of heterarchy, which defines hierarchy as complicated, fragmented, and influenced by various agents whose power varies with time, is better suited as a theoretical starting point to analyze the evolution of local governance around controversial projects with numerous agents involved. While some of these actors try to reinforce and better represent the public interest, others are out for their own private gains while others still attempt to safeguard the effectiveness of legal institutions, such as the EIA, to name but one. Such a discrepancy of interests is quite normal considering that while central governments have to discipline and balance the pursuit of energy supplies, with the ensuing environmental damages arising from a hasty economic development, local governments usually pay more attention to short-term economic growth issues. Under such circumstances, local and central interests, in regards to the implementation of environmental policies, differ. This conflict of interests between national regulations and local stakeholders has, in the past few years, become influenced by new emerging powers and actors like social organizations, journalists, legal experts and local and national agencies.

As we shall see, especially the last three of the four projects analyzed, namely the eighth naphtha cracker, the Taipei Dome and the Tamsui North Shore Road Project, comprise a mixture of successes and failures, the meeting of various levels of government, and the competing interests of the various actors.

In all of these projects different governmental bureaus and agencies often have different agendas and try to circumvent each other, a thing that has led some scholars to use the expression "iron rectangle" when referring to the interest ties between the state, the ruling party (執政黨 *zhizheng dang*), local factions and business conglomerates in Taiwan at the expense of the environment and the public interest (Kuo, 2000). Such was the case in the controversy over building a cable car in Yangmingshan National Park (陽明山國家公園 *yangmingshan guojia gongyuan*), north of Taipei. In that case, the Taipei City government tried more than once to bypass the legal requirements of the EIA and circumnavigated the Taiwan Environmental Protection Administration's (henceforth EPA) control (Su, 2006: 2). Hsu argues that in the case of watershed management, for example, democratization has had a negative effect due to the exaggerated influence of this iron rectangle in the policy process (Hsu, 2003). In reality, the problem lies in the explicit support that is being given, often without thoroughly conducted EIAs, to controversial projects, thus making "widespread corruption" and "pursuit of economic interests" the two main culprits (Grano and Tu, 2012). It has to be noted that in Taiwan the situation has increasingly improved since the landmark case of the Hsiangshan Tidal Flat Development Project in Hsinchu (1992–2001), in which an EIA for the first time managed to prevent the realization of a highly damaging developmental project. In that particular case, for the first time, a scientifically based discussion by scholars, academics and activists substituted previous emotional environmental protests (Tang, 2003).

Such overlapping competencies and increasingly complicated networks of governance that have arisen in the environmental realm in Taiwan are also visible at the global level. In Khagram and Ali's words:

> A more complex political economy is emerging globally in which the actors of governance stretch beyond the confines of states to where governmental, civil society and private sector actors form functional coalitions and broad horizontal networks, and the institutions of governance include not only international laws and treaties, but also the use of norms, codes, and voluntary standards.
>
> (Khagram and Ali, 2008: 133)

While classic policy studies tend to focus on institutional actors (governments and states) and their politics, because these are the most authoritative figures – those with the capability of turning policies into actions and regulations – lately, scholars have started to consider another angle, one that gives high importance to personal and ideological politics and that falls into the category of critical policy studies. In doing so, they take into consideration how values, opinions and ideas are shaped by the policymaking discourse and how these, in turn, shape the policymaking process (Whitman, 2008: 163). The non-state sector is one of the most active players recently emerged in governance (Zheng and Fewsmith, 2008), capable of influencing the policymaking process and the implementation of regulations at the local level.

Studies that analyse Taiwan's current development as interactions between localized traditional systems and new forms of regional governance (driven by powerful national and international forces and interests), from the point of view of various actors, are largely missing.

Tang Ching-ping and Tang Shui-yan demonstrated in one of their articles how the ruling party, in order to be successful in a confrontation with environmentalists, needs to have a strong political network at the local level, where patron–client ties are thicker and can be exploited when needed (Tang and Tang, 1997: 293). This interaction between national and local political processes is directly related to a deeper understanding of the democratization process. As argued by Diamond, "democratization inevitably proceeds unevenly, and authoritarian enclaves persist most stubbornly at the local or provincial level" (Diamond 1999: 244). Local politics are also more easily subjected to patron–client networks, which follow private interests and promote favoritism rather than transparency or accountability. Thus, public projects often become private rent-seeking activities rather than being made for the public good. While in "young democracies" central level politicians are usually committed to democratic principles (if for no better reason than simply because of their public profile), at the local level government officials still maintain patronage networks with their supporters, involving the employment of material gifts and favoritism (Tang, 2003: 1030). Traditional and long-lasting practices, dating back to the country's authoritarian era but thriving and surviving in the so-called "third wave" democracies of which Taiwan is such a perfect example, still influence the functioning of newly created political institutions (Tang and Tang, 1997: 294).

As aptly put by Shelley Rigger:

> Clientelism is a political style in which politicians (the patrons) form lasting relationships with people below them (their clients). These relationships are based on mutually beneficial exchanges. The patron gives the client access to the spoils of power, and, in turn, the client supports the patron's political ambitions and activities.
>
> (Rigger, 1999: 41)

Clientelism is a distinctive trait of politics in Taiwan, especially for what concerns the two main parties present on the political scene, the KMT and the DPP. However, when public expenditure issues are debated, citizens too become involved in this equation, joining concerned environmentalists and social groups in demanding to know *how* and *why* public money is being spent on potentially environment-damaging projects. The following question thus arises: Which actors are the most influential and how do they manage to gain support for their cause? So far, environmental sociologists have identified four categories of issue entrepreneurs as the most influential in environmental politics: media professionals, scientists, lawyers and politicians (Hannigan, 2006). The Green Party, one of the most active players present in the environmental governance scene in Taiwan, falls outside these schemes. In fact, the traditional "framing" of political campaigns used by green parties around the world, which mainly centers on spreading the party's principles of "promoting a sustainable way of life"; "increasing the number of green areas in urban centers"; and "battling pollution" has turned out to be extremely ineffective, with many inside Taiwan ignoring or simply not caring about such issues. As stated by Dafydd Fell, the top five issues in importance, for politicians in Taiwan are, for the KMT: national identity, economy, stability, prosperity and education; for the DPP, in order: national identity, money politics, democracy, social welfare and last environmental protection (Fell, 2012: 87).

In all of the cases presented throughout the book, the situation is not unitary; there is no harmony between two powerful actors (local and central government) nor is there agreement, for that matter, among the people. With this book, I follow Weller's attempt to challenge the often-used binary distinction between "central" and "local" and show that it is too simplistic in its approach, as it does not take into consideration a variety of other actors, their economic and personal interests, and the overall changes in environmental behavior that are mostly seen at the central level, where influences from abroad are felt more strongly (Weller, 2006: 169). As explained by Weller in his comparative analysis of nature perception in China and Taiwan, the most common anthropological response to the top-down model is the study of how local communities resist, rework or subvert these policies. While anthropology tends to consider the "state" a unified independent actor to be treated as a whole (in contrast to local communities, which are treated as fragmented and complex), political science tends to do the opposite: It underscores distinctions among the different levels of the state and their competing interests but ignores the differing interests among local communities, including their differences of power and culture (Weller, 2006). In this book I will try to challenge both of these views, showing that a complex interaction of state and local communities at various levels does not allow us to employ a binary dyad of "central versus local," as even within the same level of government we find different behaviors among county and municipal environmental protection authorities and sometimes different interests between the personnel of environmental and other bureaus.

This research aims at showing how both of these approaches are unsuitable to reach a full comprehension of Taiwan's multi-level and multi-stakeholder governance. During these contentious processes, the boundaries of the political sphere are redrawn, while economic and political scales are redefined. Many countries have experienced institutional voids with regard to the new challenges posed by state re-scaling. This has been evident in cases regarding the management of water and natural resources, the coordination of regional and national development, among others.

Existing scholarship has long studied societies and countries where the lack or weakness of certain institutions hinders a balanced social and economic development. Yet, until recently, scarce attention was paid to the questions of *how* and *under what circumstances* institutional voids can be filled by a multitude of new actors, present on the political scene. Whereas a growing number of studies (e.g., Khanna and Palepu, 1997) have documented intriguing cases where entrepreneurial actors – public and private, organizational and individual – have trespassed on the boundaries of formal rules and institutions to create new spaces of governance that perform important coordinating functions, there is still a lack of evidence concerning the problems of void-filling during state re-scaling and there is a great vacuum for what concerns Taiwan in particular, and the importance this country exerts, in light of its role as a potential example to be followed by China, as a democratic country with a responsible attitude towards the environment.

The growth machine theory

Business elites have a strong personal financial incentive to maintain pro-growth public policies and thus often seek alliances with local politicians who can facilitate their investments (Tang, 2001). In turn, local politicians, conditioned by the lack of resources to govern efficiently, seek alliances with those very same powerful business elites who can facilitate their rule. This vision of a city's future to be developed according to the imaginations and desires of business and political elites – a theory that is particularly apt if applied to the urban renewal/commercial development case, regarding the construction of the Taipei Dome – has been defined as the theory of the "growth machine" (Molotch, 1976, 1993). As explained by Clarence Stone in his study on Atlanta's urban politics, since every local leader has to take into account the requests and interests of a varied and fragmented electorate, business elites, with wealth and resources at their disposal, become an ideal partner for politicians, who can thus satisfy different stakeholders (Stone, 1989: 3–12). Although the growth machine theory best applies to democratic political systems with free competitive elections, research that has applied this theory with various degrees of success to one-party authoritarian systems, such as China or Taiwan in the pre-reform period, does exist. Zhu Jieming, for example, applies the concepts of growth machine and growth coalition to China's urban land reforms (Zhu, 1999). Tang Ching-ping, in his study on democratizing urban politics in Taiwan, claims that under the one-party rule, preceding the lifting of martial law, Taiwan fit the regime model well enough since even under such authoritarian and rigid constraints there were "frequent political successions through local elections" coupled with an overall continuity in pushing forward with numerous pro-growth economic projects (Tang, 2003: 1033).

Farglory Corporation and the numerous scandals surrounding its chairman and his alleged bribery of local politicians to secure developmental bids, illustrate the link that often exists between wealthy business leaders and mid-level politicians, mostly preoccupied with electoral concerns. While it is no mystery that political elites are usually more successful at enforcing regulations and pushing forward their agendas in authoritarian contexts, Taiwan incarnates a special case, being a democratic country in which, until recently, even though dissent voiced by environmentalists or concerned citizens was not stifled, such concerns were mostly voiced but led to no actual change in the dominant, pro-business agenda. This is true, as we shall see with the last two cases analyzed, because at the local level politicians are often deeply enmeshed in the economic environment; they sometimes run private businesses and thus enjoy special privileges and have a vested interest in pushing through with local developmental projects so that their businesses can profit from them. Nevertheless, the situation seems to have changed in the past few years with several developmental projects being canceled or stopped thanks to the repeated and concerted efforts of citizens from all walks of life, uniting under the common banner of fighting against injustices and abuse of authority. One of the key phenomena characterizing the newer fringe of Taiwanese civil society is the alliance of groups fighting for local issues with those fighting for broader concerns, such as obtaining governmental accountability (Cole, 2014). These coalitions, rather than simply including the traditional iron rectangle of government officials, ruling party members, business and other interest groups, include journalists, social movements, researchers and intellectuals, poets, writers, lawyers and other numerous actors, at all levels of society and government (Sabatier and Jenkins-Smith, 1993: 117–119). They are composed of a variety of actors that would never have come together if not for the "problem" they are opposing. Members of such coalitions have usually not been working together for a long time but share the same set of policy core beliefs and are able to successfully coordinate themselves because they consider the opponent as likely to cause substantial impact upon them in case of victory (Sabatier and Jenkins-Smith, 1993: 120). Events that rise to the national level have a chance of mobilizing an even greater support for the cause by involving segments of the population that while not directly engaged in such issues, nevertheless have a chance to express their views and opinions on the policy/issue in question. If the media gets involved many more people are drawn in; even those who are otherwise apathetic get confronted with such issues and, among those, some will be pushed to participate in some kind of way even if just by forming an opinion on the issue. Mass groups, composed of individuals from different backgrounds and classes, have been gaining prominence in aiding the green movement in Taiwan reaching its goals lately, for instance in the case of Taiwan's eighth naphtha cracker as well as in the case of NPP-4.

To elaborate on the above arguments, this book first depicts respectively the politics of nuclear energy, of petrochemical energy, and, in the last two cases, of downtown urban renewal and infrastructure building in northern Taiwan as these evolved into protracted contests between environmentalists on one side, and a pro-development faction composed of politicians and real estate developers on the other. Second, by delineating and analyzing the various cases, this research will show how civic associations have been able to stall the ultimate decisions of pro-growth developers and traditional power elites, for lengthy periods of time, by employing legal tools and mechanisms such as the EIA.

Environmental protection framework in Taiwan

From the early 1950s to the late 1970s, when planning authorities began considering nature as a resource that could be both exploited or nurtured, Taiwan completely lacked a system for environmental protection or nature conservation (Hsiao, 1999: 33). Solid wastes accumulation, soil, water and air pollution on a massive scale prompted planning bureaucrats to set up the framework for the environmental protection system to come. While ENGOs and individual citizens play an important role in promoting environmental awareness while pressuring autocratic regimes into compliance, as stated by Williams and Chang (2008: 74) environmental protection has to be the ultimate responsibility of the government; in fact, there is no viable substitute for public policies enacted at the central level and then enforced at the localities by the numerous environmental state agencies. It goes without saying that how the government performs, depends on the ruling party.

To better understand why the government is still struggling to accord the right place to environmental protection, it will be useful to briefly sketch the contours and main pieces of legislations that comprise Taiwan's environmental protection framework, which are fundamental for a deeper understanding and a correct interpretation of the various case studies that follow the background chapter.

Environmental protection agencies

In most countries, environmental protection and cleaning-up industrial pollution processes are the responsibilities of the government.

While citizens and activists as well as third actors can also play a crucial role in influencing rules and regulations, environmental policies are effectively enacted and enforced by governmental agencies at both the central and local levels (Williams and Chang, 2008: 90). Since Taiwan has become a multi-party system where the KMT and DPP have taken turns in governing the country, environmental protection has been, in turn, influenced by the party ruling the country in that specific moment. As previously mentioned, the two main political parties present on the scene, the DPP and the KMT, have both come under attack by pro-environment activists for being pro-business and for having betrayed their original commitment to protect the environment (in the DPP's case).

While environmental problems spurred by rapid and unchecked economic development were already evident in the 1970s and 1980s, a serious commitment on the part of governmental organs to repair the damages brought on by a fast-paced development, began in earnest after the watershed year of 1987. While a Bureau of Environmental Protection had in fact been present since 1982 under the Department of Health (DOH), it is after the lifting of martial law in 1987 that the environmental movement gained momentum and a cabinet-level agency to protect the environment, the Environmental Protection Administration (環境保護署 *huanjing baohu shu*) was formed (Williams, 1992: 200; Ho, 2006: 27). EPA, the only agency charged with protecting the environment island-wide, employs nearly a thousand people and had a yearly budget (as of 2012) at its disposal of NTD 12.42 billion.[24] This agency, which has been for many years the leading organ charged with protecting the island's fragile environment, has indeed enjoyed numerous successes and reached significant targets in the reduction of solid wastes, in spreading environmental awareness and in reducing air pollution; however, numerous problems remain yet unsolved.

As shown in Figure 1.2, EPA features a complex mix of bureaus, sub-units and local branches in charge of different tasks regarding environmental protection.

Among its numerous departments, EPA's most important divisions are:

1 The Department of Comprehensive Planning, responsible for promoting environmental education, reviewing environmental impact assessments and for enacting environmental policy.
2 The Department of Air Quality Protection and Noise Control, responsible for monitoring the quality of air and greenhouse gases emissions among other tasks.
3 The Department of Water Quality Protection, responsible for monitoring the quality of the island's water resources.

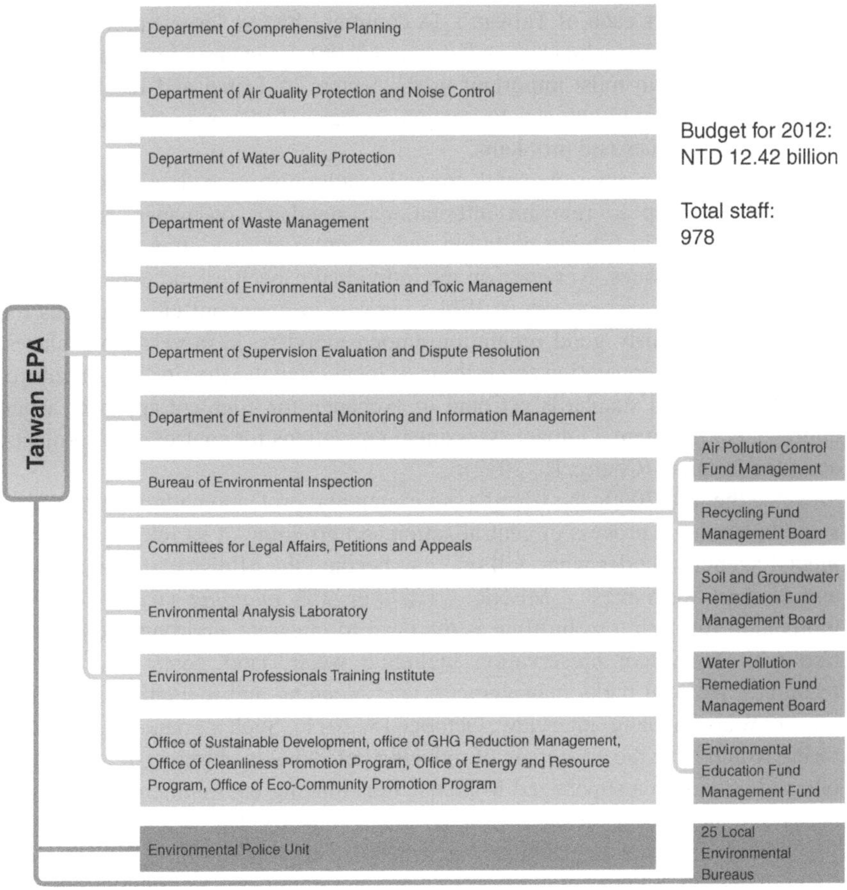

Figure 1.2 Environmental Protection Administration structure (original data source: EPA website).

While EPA is by far the most important and authoritative player in environmental protection in Taiwan, it is by no means the only agency whose competences deal with ecological or pollution issues (Williams and Chang, 2008: 92).

An equally important administrative organ is the Ministry of Economic Affairs, which has several sub-units, such as the Water Resources Bureau, the Water Conservancy Agency, the Energy Commission and the Industrial Development Bureau, all involved in several environmental issues as well. Environmental education lies instead under the responsibilities of the Environmental Protection Group under the Ministry of Education.

Taiwan's eight national parks, six on the main island and two on the outer islands (金門 *Kinmen* and 東沙 *Dongsha*), are administered by the Department of National Parks, under the Ministry of the Interior.

The Council of Agriculture is responsible for the administration of Taiwan's Forestry Bureau and Bureau of Water and Soil Conservation as well as for promoting the well-being of farmers in Taiwan.

At the localities each of Taiwan's 18 counties, the five largest municipalities of Chiayi (嘉義), Hsinchu (新竹), Keelung (基隆), Taichung (台中) and Tainan (台南) and the two most important urban centers of Taipei and Kaohsiung all have their own Environmental Protection Bureaus (EPB), responsible for local environmental issues and problems.

While Taiwan's environmental protection administrative framework is well developed and up to relevant international standards on paper, reaching a positive outcome in solving pollution and related problems often depends on a variety of other issues. Not least on the individual capacities and contacts of the personnel staffing these bureaus. While bureaus in important cities such as the capital enjoy a fairly good reputation among activists, in more distant places, stories portraying corruption and bribing of local officials who, instead of enforcing environmental standards get paid in exchange for a lenient approach when applying environmental impact assessment regulations for controversial projects, are quite frequent (Chang, R., 2012).

Since the mid-2000s the island's environmental protection administration has started planning a process of centralization and grouping of its relevant environmental agencies, under what will one day become the Ministry of Environment and Natural Resources – MENR – (環境資源部 *huanjing ziyuan bu*). The motivation for such a reshuffling is the need to integrate environmental protection with ecosystem preservation including water, land, forest and mineral resources, national parks management, land planning and regulation of nuclear power (Winston Dang, interview, October 18, 2011). Such a grand-scale restructuring would have the advantage of reducing personnel by merging several councils and units into a supersized Ministry, thus making environmental protection easier to manage. The idea of setting up such a ministry dates back at least to the year 2006, when the Legislature was scheduled to vote the bill (Williams and Chang, 2008: 94). Political instabilities have so far prevented it from becoming a reality, even though there is speculation that the reshuffling might take place in 2015 (Chu Yu-chi, interview, November 2, 2011).[25] The rationale behind this centralization is the unification of different responsibilities under one agency, in order to solve previous numerous impasses and inefficiencies caused by an unclear and inefficient distribution of environmental responsibilities and tasks among numerous department and agencies with conflicting agendas. The MENR will mainly deal with issues that can be categorized under three broad categories: *Pollution Control*, *Resource Use Planning* and *Nature Conservation* and, once established, it will become Taiwan's largest ministry, bringing thousands of governmental employees under one roof.

Environmentalists in Taiwan have been long-time supporters of the creation of a ministry of the environment, which they believe will improve environmental policies' integration into economic planning and facilitate the actual enforcement of environmental legislation (Peter Wushou Chang, informal communication, August 31, 2011; Hsü Kuang-jung, interview, September 15, 2011; Tsui Shu-hsin, interview, September 20, 2011).

As aptly put by Williams and Chang (2008: 95), successes or failures of EPA so far, have depended mostly on its leadership, which must possess certain attributes (good skills in communication, coordination and cooperation) as well as be experienced in environmental issues and familiar with academic life. Anywhere in the world, but especially in developing countries where priority lies with industrial development at all costs, balancing and integrating economic growth with environmental protection can become a difficult task. Several previous EPA leaders have taken their role as safe-guarders of the environment seriously and a few of them have resigned upon realizing the difficulties in trying to reconcile central directives for coastal industrial parks, tidal land reclamation and other pro-development projects with environmental protection.

The Environmental Protection Agency's top management must therefore possess several characteristics, ranging from good charisma and communication skills, to technical expertise (environmental, economic and engineering) and, most importantly, a political flair to help him/her navigate between the conflicting interests of the various actors, while at the same time gaining the general public's trust (Winston Dang, interview, October 18, 2011).

Some of the past ministers have performed their job better than others. Among the most famous figures, Lin Jun-yi Edgar (林俊義), an American trained biologist, regarded in Taiwan as the father of the anti-nuclear movement (Ho, 2003: 690) was EPA's minister for one year (2000–2001). He was appointed at a very controversial time, just when the DPP first came to power and the issue of NPP-4 became, once more, a hotly debated topic. Lin's career as EPA minister ended abruptly after only one year in office, following an oil spill, which devastated a highly diverse ecological conservation area in southern Taiwan for which EPA's intervention was dubbed as "inefficient and slow" (Chen, 2004). Rumor has it though, that his anti-nuclear stance was considered "too radical" even by the "anti-nuclear" DPP (Williams and Chang, 2008: 97) and thus President Chen had to remove Lin from office, especially in virtue of the decision to go ahead with NPP-4, reached in 2001; more on this will be said in Chapter 3.

Lin's successor was Hau Lung-bin (郝龍斌), Taipei's mayor (2010–2014); Hau is a mainlander and a powerful figure inside KMT circles. He is the son of the ex-premier, General Hau Pei-tsun (郝柏村). With a background in agricultural chemistry he was quite liked by the public but criticized among more radical environmental circles for his support of nuclear energy. He resigned over a dispute regarding a citizens' referendum running counter the decisions of the EIA committee (環境影響評估制度委員會 *huanjing yingxiang pinggu zhidu weiyuanhui*) in 2003.

From 2005 until 2007 it was Chang Kuo-lung's (張國龍) turn, an environmental activist with a strong anti-nuclear background. Under his leadership fewer EIAs were passed and cooperation with ENGOs and environmental protection organizations, such as TEPU, increased, which sometimes led to strong disagreements with the central government (Williams and Chang, 2008: 99). After Chang, a DPP legislator took the post until a suitable replacement could be found. The last minister of EPA under Chen Shui-bian's administration, was Winston Dang (陳重信) who remained in charge for a little less than one year (2007–2008) and with whom I conducted several interviews during my stay in Taiwan. A former citizen of the United States, he renounced US citizenship to take up the position. Mr. Dang is a cultivated and soft-spoken gentleman, sympathetic to the plight of the environmental movement, and very pleasant to talk to.

In 2008, as Ma Ying-jeou won the Presidency, Dang was substituted by Shen Shu-hung Stephen (沈世宏), a controversial figure among environmentalists. While in office he was often criticized for employing ad hoc "anti-activists measures" – such as forcing his underlings to work overtime to write up scathing articles criticizing environmental activists and scholars (Wu Kuen-yuh, interview, September 20, 2011) – and polarizing the environmental debate along political lines (Lai Fen-lan, interview, June 29, 2011). Furthermore, many of my sources think of him as being heavily influenced by the business and developmental community. Shen is the longest serving EPA minister of all time, lasting more than five years in the post. He served as minister until March 2014 when Wei Kuo-yen (魏國彥), a professor of geoscience with a Ph.D. degree in oceanography from the University of Rhode Island, took the post. After becoming minister Wei declared that he hopes to help facilitate EPA's planned transformation into the Ministry of Environment and Natural Resources as soon as possible.

Environmental protection laws and regulations

The most essential piece of legislation enacted by EPA is the *Basic Environment Act*, which contains the fundamental ideals on which several interest groups, from concerned environmentalists to common citizens and pro-development actors, should be in agreement. This landmark legislation has a long gestational history; in fact, it was first discussed in 1988 and finally passed in December 2002 after a third reading, due to the multitude of varying opinions on it, which for years prevented the reaching of an agreement among all relevant stakeholders. The *Basic Environment Act* transformed environmental protection into a prime consideration of economic development for the first time in Taiwan's history.

It is worth noting that one of the fundamental tenets of the *Basic Environment Act* is the commitment to render Taiwan a nuclear-free country, which is in line with the DPP's traditional position on the matter, an issue that will be dealt with in the next chapter.[26]

Basic provisions of the law require all levels of government to commit to curbing air-pollution emissions and discharge of waste waters (a provision that, as we shall see in the case of Taiwan's eighth petrochemical complex project, is not always observed, especially at the localities). Under this law Non-Profit Organizations (NPOs) are allowed to file a lawsuit against the government in case the latter is presumed guilty of negligent or damaging behavior towards the environment (Williams and Chang, 2008: 101). While this last arrangement was considered as revolutionary at the time when the law was proposed in 1988 (after 40 years of authoritarian martial-law regime), contemporary Taiwan has become accustomed to accommodate the numerous claims of a vibrant and outspoken civil society, which battles various issues through the employment of legal regulations and participatory approaches, in a dynamic environmental governance system.

Another important provision pertaining to the Basic Act is the donation, on the part of the state, of monetary funds to NGOs and NPOs, which dedicate their efforts to spreading environmental education and awareness. While this has undoubtedly turned out to be one of EPA's biggest successes, it has also had the secondary effect of turning some of the organizations that have become target donors of such endowments, into less aggressive and more subservient entities. This is not to say that all organizations that have received governmental funding have become less confrontational towards the relevant authorities, but those who are collaborating with the government and chose to abide by an unwritten rule of "respectful cooperation" make no secret of it.

On a rainy and sleepy afternoon, so typical for Taipei in the springtime, Mr. Lai Rong-xiao (賴榮孝), chairman of the *Society of Wilderness*, was taking me back to *Ximending* metro station, after our interview when, en route, we stopped in front of EPA to let out some of his colleagues. After they exited the car I asked Mr. Lai whether *Huangye* (as the group is known inside Taiwan) participated in the numerous protests and night vigils in front of EPA that were taking place, almost on a daily basis, in that period, such as one the previous week to protect farmers' rights. His answer was very straightforward. He said that it had occasionally happened, but that *Huangye* prefers to keep a non-antagonist and collaborative profile towards governmental authorities, following the belief that they have more to gain by acting as responsible players (Lai Rong-xiao, informal communication, September 1, 2011).[27]

While *Huangye* and TEPU are both among the organizations that have been fortunate enough to receive financial donations from the government (Hsü

Kuang-jung, interview, September 15, 2011), TEPU doesn't shy away from fierce critique towards governmental authorities, especially on the issue of nuclear energy and the island's fourth nuclear facility.

Numerous other laws compose the backbone of Taiwan's environmental legislative framework; first and foremost the already mentioned *Environmental Impact Assessment Act*, which is the subject of the next paragraph; the *Air Pollution Control Act*; the *Water Pollution Control Act*; the *Noise Pollution Control Act*; the *Waste Disposal Act*; the *Soil and Groundwater Pollution Remediation Act*; the *Marine Pollution Control Act*; the *Drinking Water Management Statutes*; the *Environmental Agent Management Act*; the *Environmental Dispute Settlement Act*; the *Waste Disposal Act*; the *Toxic Chemical Substances Act*; furthermore, several statutes regarding EPA's inspection organizations and other statues concerning the training of environmental personnel, are in place.

Within these laws, EPA has enacted several regulations, some of which have been amended several times. It is of crucial importance to emphasize the significant difference that exists between a *law* and a *regulation*. While laws are always passed by the Legislative Yuan (立法院 *lifa yuan*), sometimes referred to as a "quasi parliamentary body," although in reality one of the five branches of government defined in the constitution of the Republic of China – regulations can be passed only by EPA and are details pertaining to higher laws.

Environmental Impact Assessment Act

Environmental impact assessments are meant to minimize and forecast the negative impact of a developmental project on the ecological, social, cultural and economic environment in a certain locality.

Taiwan's EIA Act was formally enacted in 1994, modeled after the United States' own EIA Law. However, its origins go way back and the history of its implementation is long and controversial. In 1973 the United States' EIA system, put forth in America in 1969, was formally introduced by the Taiwanese government with a governmental document (Ho, 2004: 239). Initially, the motives for introducing an EIA system in Taiwan were economical. In fact, nearly all of the early EIA processes turned out to be mere showcases proffered to help bureaucrats avoid difficulties during the construction phases. The introduction of such a system was a direct consequence of the socio-political situation the country was undergoing: bureaucrats needed a showcase to appease popular requests for increased participation. A telltale sign of how insignificant the EIA's initial role was can be found in the fact that numerous projects were already approved before an EIA was actually carried out.

The Department of Health, which carried the responsibility for environmental protection until 1987, proposed a first draft for the EIA law in 1983 but due to fierce opposition by economic actors and bureaucrats, the process was long delayed. Starting from 1985 14 developmental projects underwent an EIA although, as pilot cases, their purpose was merely "educational," in the sense of serving the goal of training experts (Ho, 2004: 240). At the end of the decade the top-down façade manner in which the EIA had been conducted until that moment, began crumbling under rising popular criticism. Several secondary economic and other agencies stepped up their requests to conduct counter EIAs, with bureaucrats worrying about external interference, in what was previously their undisputed domain.

In 1990 the Executive Yuan (行政院 *xingzheng yuan*) allowed the recently established EPA (1987) to put forth a second EIA law draft. Towards the end of that same year the draft bill was finally introduced to the Legislative Yuan but it took four years to complete this legislative process, so that the EIA de facto became a law only in December of 1994. More than a decade had to pass for its codification to take place. This delay was due to several issues ranging from aggressive environmentalism weakening institutional support for the law to the opposing interests of the industrial sector and of the pro-development faction both within governmental echelons as well as among business corporations. Conflicts between junior ministers in favor of an EIA (the Department of Health and EPA) – pitted against the far more powerful and senior ministries such as the Ministry of Economic Affairs – hampered a speedy implementation and codification of the law.

Especially during Hau Pei-tsun's premiership (1990–1993) crack-downs towards environmental activists were frequent and violent and the country was not yet ripe for the passage of a law, considered increasingly as "in favour of environmental activists."[28] Besides, as the first truly representative parliament of the post-war years was elected in Taiwan, after the KMT allowed for a fully-franchised election in 1992, the DPP won more than one-third of seats in the Legislative Yuan (Ho, 2014a: 7). The opposition party's links with environmental activists made the process for a real codification of the EIA law much easier.

But what are exactly the legal provisions contained in an Environmental Impact Assessment and how does this law work?

In practice, the EIA stipulates how projects should be planned and reviewed, which requirements should be met, and what are the punishments and fines for those who violate the said procedures. The top body, which reaches the final decision in regards to the approval or refusal of a project, is the EIA committee with its 21 members; among these, seven are government appointees (機關代表委員 *jiguan daibiao weiyuan*) and 14 from academia; these are nominated by a mixed committee made of members of both the government and the academia (Winston Dang, informal communication, August 31, 2011). The committee meets once each month and the minimum quorum of members who have to be present for a session to have legal validity is 11. For the sake of clarity it has to be said that the EIA committee only gets involved in those cases which deal with projects that involve parcels of land that are larger than ten acres; smaller projects are handled by local EPBs, although some exceptions can be made, as we shall see in the case of the Tamsui North Shore Road Project, analyzed in Chapter 6 of this book.

While the EIA was first wished for by certain sectors of the state, who believed it could be useful in preventing and avoiding embarrassing street protests by re-channeling people's grievances inside the legal system (Ho, 2004: 239), leading the population to believe that thorny developmental issues were finally being settled via scientific investigations, in the past two decades things have improved.

Speculation remains as to whether the EIA system works as it should or, as some of my informants claim, it is still an easily corruptible process, tilted towards pleasing tycoons and developers (Prof. Winston Dang, interview, October 18, 2011).

In fact, although it is generally believed that the EIA system works quite well, with controversies and scandals marring its process mostly relegated to the early years of its implementation, when businesses deliberately ignored its basic requirements, all the case studies analyzed in the book will show that serious problems remain, and that accusations of bribery, corruption and of less-than-satisfactory implementation of the EIA Act on part of the pro-environment camp, are still common.

Taiwan Land Expropriation Act

When the Nationalist government fled to Taiwan, bringing along half a million refugees between 1947 and 1949, much housing and land was confiscated from both aboriginal people and earlier Chinese settlers. Land regulation and registration was carried out according to laws that had been conceived on the mainland but had never been in use until that moment. Many believe that the KMT's loss of political power on the mainland had been in direct correlation with its incapability of reforming the land tenure system, a failure that won many sympathies to Mao's Communist Party (Rigger, 2011: 46).

In the first phase of the land reform, which began in 1949 in Taiwan, the state ordered a reduction of farm rents from 50 percent of the harvest to 37.5 percent (Rigger, 2011: 46) and later on a redistribution of formerly Japanese-owned lands (Fell, 2012: 17). This move proved to be a boost for several farming families. The second phase of the land reform began in 1951 and was based upon the principle that landlords with excessive land property (any parcel in excess of 2.9 hectares) would have to sell their extra parcels to the government, in exchange for allocation of staple foods, bonds and shares in four large publicly owned companies, Taiwan Cement, Taiwan Pulp and Paper, Taiwan Industrial and Mining and Taiwan Agricultural and Forestry Development Corporations (Bain, 1993: 32). These parcels would then be re-sold to the landlords' tenants, through a 30-year loan at a low interest rate. Most Taiwanese welcomed the KMT-led land reform, which launched the country's agricultural productivity. In fact, the farming population rose from 38 percent to 60 percent in less than a decade (1949–1957) (Fell, 2012: 17). Nevertheless the ensuing confusion following the somewhat arbitrary division and confiscation of land property rights, has characterized developmental projects for years, so that the process of zoning and confiscation of land has become tangled in economic and political intrigue for decades (Arrigo *et al.*, 1996) and remains a thorny issue until today.

The new-found prosperity of the export-oriented policies in the 1970s led to a building spree all throughout the 1980s, as development plans came under public criticism for being short-term oriented and aimed at increasing land speculators' and developers' assets, through under-the-table deals.

Forced land expropriations in several townships of Taiwan in the past few years have brought some thousand farmers to the streets, demanding an amendment of the *Land Expropriation Act* (土地徵收條例 *tudi zhengshou tiaoli*) (Xin, 2013). The dispute in Dapu (大埔), Miaoli County (苗栗縣)[29] represents the most high-profile case of resistance to "zone expropriations" that has taken place in Taiwan in the last decade zone expropriations take place when large parcels of land become subject to compulsory sale to the government, for projects that use part of the land for infrastructure and sell other portions to raise funds for construction or local government's financing.

Taiwan Rural Front's spokesperson, the above-mentioned NGO whose main scope is fighting unjust governmental land appropriation on behalf of farmers and their rights, revealed in one of our conversations that there are currently 95 similar cases of zone expropriation pending and that many large-scale expropriations of farmland have taken place in the past few years, under the pretext of creating new towns or industrial zones, without any real consideration of actual needs, thinking only of land speculation deals (Tsai Pei-hui, interview, September 22, 2011). Taiwan Rural Front has, for many years, been involved in a battle with the state to amend the *Land Expropriation Act*; besides prohibiting further use of zone expropriations, draft revisions to the Act proposed by TRF and a team of lawyers would mandate that the purpose of the law is to "protect the people's right of survival and property rights"; furthermore the amendments would also mandate substantive assurance of public interest and necessity and democratic and transparent public review (Chan Shun-kuei, email communication, January 23, 2014). However, the TRF-sponsored draft revision of the act did not even get past the procedure committee of the national legislature. Again according to Taiwan Rural Front's spokesperson:

> Current legislation gives local residents and landowners no effective right to oppose land expropriation at all. Despite legal requirements for public hearings, price negotiation between the government and landowners as well as reviews by planning committees, these safeguards have remained mere written words.
>
> (Tsai Pei-hui, informal communication, October 15, 2011)

Environmentalists have been on a war footing since 2013 as rumors began circulating about a governmental plan for easing regulations on land use, titled *National Regional Plan*; if implemented, activists believe that this plan would ease restrictions on development in reservoir areas, farmland and active fault zones (Wang and Lee, 2013b: 4) thus making it easier for projects of a certain size to be approved. The application and review procedures for projects such as nuclear waste storage sites, nuclear power plants as well as casinos and other such sized facilities would also be regulated by this plan. Environmental groups, staunchly opposed, argued that this piece of regulation should not be approved without holding public hearings on it and that, as it is, it would be beneficial for big corporations only.

This claim for people to be more involved in the key decisions concerning the country is often heralded to attack the ruling party and what activists believe to be an authoritarian stance in denying the public access to policymaking by holding closed-door meetings to discuss issues, ranging from the revision of important laws on behalf of the environment, to the island's energetic future or its political independence.

The crucial point is whether the vibrant activism of the past few years will result in a cleaner environment for Taiwan and whether the visibility that controversial projects are given, thanks to activists and media intervention, will bring about positive structural changes rather than superficial, cosmetic adjustments meant to stifle opposition without sacrificing economic development, as has been the case in most instances until now.

In view of the large number of developmental projects the government is planning for the near future, which – if carried out – would expropriate thousands of acres of land and force thousands of families to relocate somewhere else, such issues are of extreme importance since they pose threats not only to the ecological environment but also for social stability.[30]

Notes

1 See also an open letter signed by several influential professors and scholars of Taiwan in 2009 sent to the *Taipei Times* on November 13: www.taipeitimes.com/News/editorials/archives/2009/11/13/2003458289 (last accessed on February 8, 2015).

2 The Wild Strawberry Movement was a series of protests by 400 students, which took place in November 2008, as a result of harsh police treatment of demonstrators protesting against the visit to Taiwan by PRC envoy Chen Yunlin (陈云林). The visit is credited for setting the stage for the signing two years later, in 2010, of the Economic Cooperation Framework Agreement (ECFA). Among other things, students asked for the abolishment or revision of the *Assembly and Parade Act* (集會遊行法 *jihui youxing fa*) to safeguard free speech and other basic human rights. The criticism is that, as it is now, the act can be used by the government to deter activists. For example, tree hugger Pan Han-chiang (潘翰疆) has been accused of violating the act after he attempted – without permission – to prevent authorities from chopping down old trees in a park to clear the way for the construction of a police station (Pan Han-sheng, email communication, June 29, 2014).

3 To name but two of the most significant ones, the "Jasmine Revolution" in China and the even more famous "Arab Spring," which began in December 2010 in Tunisia and quickly spread to several other countries (Oman, Yemen, Syria, Morocco and Egypt) during the course of 2011.

4 The *Xinhai* Revolution (辛亥革命 *xinhai geming*), which overthrew China's last imperial dynasty, the Qing, had its turning point in the Wuchang uprising, on October 10, 1911 and is commemorated in Taiwan as the "double 10 anniversary" being also the Republic of China's (ROC) national day. In 1912 Sun Yat-sen founded the ROC on Mainland China, which governed various parts of the Mainland's territory until 1949. When communist forces prevailed, the ROC government fled to Taiwan. In

1971 the People's Republic of China (PRC) assumed China's seat at the United Nations and the ROC, which until the 1990s still officially claimed to represent all of China, is currently recognized by only 23 countries in the world (Fell, 2012: 158).

5 Lee Chao-shing (李昭興), a professor of applied geosciences at National Taiwan Ocean University, was quoted in the *Taipei Times* saying that as many as 70 underwater volcanoes are located within an 80km radius of the fourth nuclear power plant in *Gongliao* District (貢寮區), in New Taipei City (新北市 *xin beishi*). Up to 11 of those volcanoes are active, according to Lee (Chao, 2012).

6 Taiwan Power Company (台灣電力公司 *Taiwan Dianli Gongsi*, also known as "Taipower") is a state-owned electric power company providing electric power to Taiwan and to the offshore islands of the ROC.

7 On May 18, 2011 the water supply for ordinary, non-industrial households in New Taipei City's Linkou District, Taoyuan County, Hsinchu County and Hsinchu City was curtailed by 20 percent, while industrial users had to deal with a 5 percent cut in their supply. The same measures became effective for Miaoli County, Taichung City, northern Changhua, Tainan City and Kaohsiung on May 23 (Mo, 2011b).

8 Unfortunately, several other food-safety scandals have since then plagued Taiwan in a crescendo of devastating events; the most (in)famous one being the tainted oil issue, which exploded in September 2014 (*The New York Times*, 2014).

9 The ECFA, signed in 2010 between China and Taiwan, provides a basis for further integration of the two economies and represents a milestone in institutionalizing cross-strait relations. In Taiwan, opposition politicians continue to criticize the ECFA but months of discussion and debate have led to a gradual increase in public support and acceptance. The government, on its website (www.ecfa.org.tw), has referred to it as the precedent of the Cross-Strait Service Trade Agreement, which came under public scrutiny and criticism in 2014 and which is credited for giving rise to the Sunflower Student Protest. Official translation of the ECFA can be found at: www.ecfa.org.tw/EcfaAttachment/ECFADoc/ECFA.pdf (last accessed on November 30, 2014).

10 The Sunflower Student Movement (March 18–April 10, 2014) is a protest movement by students and civic groups, which occupied the Legislative Yuan and, later, also the Executive Yuan of the Republic of China. Students protested the passing of the Cross-Strait Service Trade Agreement (CSSTA) by the ruling Kuomintang without previously carrying out a clause-by-clause review. The CSSTA had already been negotiated and signed in mid-2013 by the two organizations that represent Taiwan and Mainland China in cross-strait relations, the Straits Exchange Foundation (海峽交流基金會 *haixia jiaoliu jijinhui*) and the Association for Relations Across the Taiwan Straits (海峡两岸关系协会 *haixia lianganguanxi xiehui*); CSSTA is actually the follow-up to the already mentioned ECFA, which had been in place since 2010 and had removed trade restrictions between the two sides of the Taiwan Strait on a wide selection of products. The CSSTA covers the service sector. The choice of associating the movement to sunflowers was apparently made after a florist delivered bunches of sunflowers to the students to symbolize the need to let some sunshine in on the government's behind-the-curtain deals. Clearly, it also is a reference to the Wild Lily Movement of 1990. Furthermore, sunflowers had already been used in past antinuclear protests because, in virtue of their heliotropism (meaning that they move by tracking the sun) they are perfect to symbolize the hope for clean and renewable energies.

11 The Municipal or "Nine-in-one" (九合一選舉 *qiheyi xuanju*) elections were held on November 29, 2014, for all directly elected local governments' positions, such as the mayors and councillors of the special municipalities; county commissioners and councillors; city mayors and city councillors; township administrators and councillors; and finally borough village wardens. This election was so far the biggest in Taiwan's history in terms of the number of representative posts up for election, with a record-high of 11,130 public servants to be chosen for nine categories of elected offices. After the KMT

suffered a massive setback in the polls, which took many in the party and among the public by surprise, speculations as to whether Ma should resign as party chairman and as to whether Taiwan's China-policy should continue as it is, began in what was seen as a referendum on Ma Ying-jeou's presidency so far. Finally, Ma quit his post as KMT chairman in December 2014; this was then assigned to Erich Chu.

12 For an explanation of the "China Outside" concept, see Shelley Rigger's *Why Taiwan Matters* (2011: 133).

13 EJA was founded by Robin Winkler and other lawyers and scholars dedicated to environmental issues in 2010. EJA was involved in several major cases, such as the Tamsui North Shore Road Project (淡北道路 *danbei daolu*), which is the subject of Chapter 6 in this book, NPP-1 and the third expansion phase of the Central Taiwan Industrial Park (中科三期 *zhongke sanqi*).

14 The *Society of Wilderness* is an organization dedicated to preserving nature and wilderness in their original state, protecting them from pollution and human destruction. Among their activities are: fundraising for the purchase of pristine land, promoting awareness of nature conservation, organizing natural education courses and providing counselling and assistance to governmental agencies dedicated to protecting wilderness and natural resources.

15 *Taiwan Environmental Protection Union* (TEPU) is Taiwan's oldest (founded in 1987) and leading environmental protection organization with over a thousand members recruited among university professors, students, housewives, writers, filmmakers and many other people, from all walks of life, who are concerned about Taiwan's environment and who witnessed the early abuses that the state-led industrialization of the 1960s and 1970s brought upon their country. For more details see: http://tepu.yam.org.tw/intro.html (last accessed on February 8, 2015).

16 *Taiwan Rural Front* (TRF) was established in 2008 by a group of farmers, academics and students opposed to the government-backed *Rural Revitalization Act* (農村再生條例); this organization, aimed at protecting farmers' rights, was initially active in promoting and designing seminars on rural communities and on farmers' livelihood and needs. TRF has meanwhile evolved into a social-justice group dedicated to working with victims of the government's forced land seizures while at the same time pushing for legal and policy reforms. Taiwan Rural Front's current activities include holding meetings in rural villages nationwide, advocating to put a stop on unjust appropriation of farmers' land by the government for industrial development purposes. This group is characterized by a strong anti-globalization position and is firmly opposed to the various trade pacts signed with the PRC since 2010. In fact, its philosophy maintains that land and labor should not be treated as mere commodities but should serve the greater good of empowering individuals by giving them a purpose; in the case of farmers this means doing something good for nature by not using pesticides, while the Taiwanese public can choose to buy such natural products over the ones produced by profit-driven corporations. For more information see: www.taiwanruralfront.org/ (in Chinese).

17 The *Mothers' Alliance for Monitoring Nuclear Power Plants* was established in March 2013 by Chen Ailing, who acts as the chairperson of the Fubon Financial Group's charity foundation; her husband is the chairperson of the Fubon Financial Group, which maintains close ties with Ma Ying-jeou. Before taking up the role of activist, Chen met with Tsui Shu-hsin, to enquire how to steer clear from the "KMT-DPP partisan rivalry" (as quoted in Ho, 2014a: 17).

18 The *No Nuke Cultural Activism Group* was established in 2009, in protest against Taiwan's fourth nuclear power plant, in Gongliao. In less than one year the group, which started from a loose, informal association, set up a music band named *No Nuke Street Band* (諾努客走唱隊 *nuo nuke zou chang dui*). Through its music, this band is much better suited for communicating with younger generations, getting them interested in environmental and energy issues, which were previously not so appealing for Taiwan's youth.

19 The *Organization of Urban Re-s* was established by a few urban residents of Xinyi District, in order to fight the plans of developers, business corporations and the city government for the development of the real estate sector of what was rapidly turning into Taipei's up-scale financial district. While this group's actions were initially localized and directed at Xinyi, their focus has gradually expanded to fighting for a correct and respectful use and planning of Taipei's urban environment. For more information see: www.ours.org.tw (in Chinese).

20 The *Homemakers United Foundation* was set up in January 1987 – a few months before martial law was lifted in July – although it officially became a foundation only two years later, in 1989. In the words of the group's chairwoman, Chen Man-li (陳曼麗), with whom I met several times during my stay in Taiwan, among their main activities Homemakers is dedicated to supporting Taiwanese farmers and local producers over imported products while at the same time being vehemently opposed to Genetically Modified Organisms (GMO). At the early stages of its activism, it was among the pioneering organizations that pressured the government into starting a grand-scale, national recycling campaign. It successfully did so, by organizing several "trash-recycle movements" (垃圾分類活動 *lese fenlei huodong*) in front of EPA all throughout the early 1990s (Chen Man-li, interview, July 20, 2011). The spirit of the foundation, at the time when it was set up, was to show that women could be something other than merely housewives, caretakers and cooks and that they too had an important role to play in society outside of the family.

21 The *Wild at Heart Legal Defense Association* is a public interest organization dedicated to promoting efforts to restore a healthy relationship with nature; established in August 2003 it tries to conserve and preserve Taiwan's fragile environment. Among its professional activities WAH offers legal assistance and consultations to environmental groups or concerned individuals who fight on behalf of nature but lack sufficient funds or legal knowledge to fight by themselves. Among their most prominent members is US attorney Robin Winkler who ran for the Green Party in the presidential election of 2012. In the end the Green Party did not manage to reach the 5 percent share of votes that would have granted them a seat in the parliament (Chiu, 2012). For more information on WAH see: http://en.wildatheart.org.tw/archives/about_wild_at_heart.html (last accessed on February 8, 2015).

22 "*Primordial Law Firm*, established in 1996, is engaged across a broad spectrum of legal services, such as international trade, intellectual property, banking, merger and acquisition, commercial arbitration, administrative procedures and environmental law" (as quoted in: www.justuslaw.com.tw/eng/profile.php, last accessed on February 8, 2015).

23 Due to the recent changes that have taken place in regards to this project, starting from the summer of 2013, the chapter was updated and revised in June 2014. The final version is thus quite different than the original article published in the *Journal of Current Chinese Affairs* in 2012 and co-authored with Tu Ping-lan.

24 Source: Introduction to the Environmental Protection Administration, April 2012: www.epa.gov.tw/en/artshow.aspx?art=20071224131905l1&path=9103&list=9044 (last accessed on November 22, 2014).

25 The draft act for the upgrading of EPA into the Ministry of Environment and Natural Resources, plus draft bills for its subordinate organizations, was sent to the full legislature for review in May 2013. As of August 2014 the review process has been put on hold due to irreconcilable differences between the ruling KMT and the opposition parties, DPP and Taiwan Solidarity Union (TSU).

26 The commitment of DPP towards the environment in general and more specifically towards the thorny issue of the fourth nuclear power facility is questioned by environmental activists on the basis of the fact that when the DPP had the power to scrap NPP-4's project, under Chen Shui-bian's presidency, it chose not to do so in order to safeguard economic imperatives (Williams and Chang, 2008: 118). More about this issue will be said in Chapter 3.

27 Spontaneity being one of the basic tenets of civil society, one might wonder as to whether collaborative social movements are still to be considered as belonging to this category.

28 Hau's premiership was particularly harsh for activists and social movements due to his radical categorization of environmental campaigners as "hoodlums," which, in his words, could be of three kinds: "environmental, labour and agricultural" (Tang and Tang, 2004: 181).

29 The sadly famous Dapu expropriation case involved four homeowners who became victims of land expropriation on July 18, 2013, as their houses were torn down by the local government, to make way for a science park. The controversy was made worse by the fact that their houses were demolished while the owners were away pleading with the central government to stop the demolition. Police and governmental behavior triggered spontaneous civil disobedience protests by ordinary citizens across Taiwan. On January 3, 2014, a panel of three Taichung High Administrative Court judges found that the compulsory purchase of land belonging to Ms. Peng Hsiu-chun, Mr. Chang Sen-wen's widow (Chang Sen-wen was found dead under a bridge in what seemed to be a suicide in July 2013), and eight other citizens in four households and the July 18 demolitions were "illegal" and voided the expropriation order approved by the Ministry of the Interior (MOI).

30 One such case is the Taoyuan Aereotropolis (桃園航空城 *taoyuan hangkong cheng*), a project that is meant to expand the current capacity of the country's first airport; the project is supported by the local Taoyuan County government and will need a total of 3,200 hectares of land with over 40,000 people having to be relocated in order to expand the Taoyuan airport and build a third runway.

2 Contextualizing Taiwan's environmental history

From "developmentalism" to ecological awareness?

Early environmental depletion

Environmental deterioration goes way back in Taiwan's history reaching as far back as Qing times; a period in which the first mass migrations from mainland China brought increasing distress to the island's fragile forest cover and its coveted camphor tree (Williams and Chang, 2008: 14). Later on Japanese rule (1895–1945) also contributed to worsen the ecological situation by initiating the process of dam building (Kerr, 1965), whose serious side effects are still visible today and have, in recent years, come under intense scrutiny and criticism from the environmental community. While the Japanese era was all about maximizing the island's potential for serving as a natural resources base for the benefits of the Japanese Empire, at the same they tried to avoid any reckless exploitation of its forest covers, water and resources, wanting to turn the island into a model colony with long-term benefits for Japan (Williams and Chang, 2008: 16).

From 1947 onward the Chiang Kai-shek era had devastating consequences on the environment (Kerr, 1974; Liu, 1987). Initially, after the Japanese lost the war only a reduced number of administrators and troops were sent to Taiwan; however, after the communists managed to conquer the mainland, the ROC government and the remainder of the nationalists' troops, more or less two million individuals, fled to the island. While many newcomers settled in the north, around the Taipei area and Hualien area, where they took over the structures left behind by the Japanese (Rigger, 2011: 31), earlier Chinese immigrants, who had been brought to Taiwan by the Dutch, had settled primarily in the South and central areas and in the remote wilderness surrounding Ilan (Rigger, 2011: 62). This division, based on ethnicity and political orientations, has continued to divide the political life of the country until today with far-reaching implications and influences on numerous contemporary social issues, none the least the island's fragile environment (Williams and Chang, 2008: 18).

Environmental depletion under Chiang Kai-shek

This period marks the beginning of the Republic of China being established in Taiwan, which is divided in two periods, the martial law era (1947–1987) and the post-martial law era (1987 until today). During the 40 years of martial law, political dissent and public participation in developmental plans were almost non-existent and ecological awareness yet a distant concept.

While the Chiang Kai-shek government was initially preoccupied with reclaiming its rightful position as ruler of both Taiwan and the mainland, its economic and developmental strategy betrayed its position vis-à-vis the island, viewed as a mere temporary retreat. Under this operational framework, natural resources' exploitation had to be maximized and subjected to the primary goal of rescuing China from the communists (Arrigo *et al.*, 1996: 767). Roads, railways and hydroelectric infrastructure initiated during the Japanese era were further expanded through the construction of cross-island highways and smaller roads for facilitating logging and mountain resources' exploitation. All of these projects were carried out without consulting political opponents or listening to people's concerns; the highest precedence was accorded to defense and economic development and environmental distresses faded into the background. At that time Taiwan's authoritarian government created a classical situation of governmental decisions reigning free without a system of effective checks and balances on its pronouncements; all the while the attitude was of total disregard towards both public welfare and the ecological situation. Since the KMT's arrival on the island, the authorities' main concern was a thriving economic development; at first for security reasons (the China threat), and secondarily for internal reasons (need to grow in competitiveness). For both the government and its citizens (mostly preocçupied with their economic well-being) environmental issues remained for a long time mere secondary troubles, buried under the powerful rhetorical imperative of "reclaiming the mainland" (Tang, 1997: 283; Williams and Chang, 2008: 1). As the idea of reclaiming the mainland gradually faded away, due to the changing international recognition of the PRC as the "real China," the KMT finally turned its attention inward, focusing on the development of an export-oriented economy.

In the 1960s, under the protective umbrella of the United States and its consumer market, Taiwan's small and medium-sized export industry grew exponentially (Williams and Chang, 2008: 20). Lack of laws and regulation as well as faulty plans for zoning and control of various industries (Chen and Shih, 2010: 19) led to a dispersion of industrial wastes and pollution problems all over the island's territory. To divert attention from political problems and regain sympathies among the native Taiwanese, the KMT promoted rural industrialization with thousands of small and medium-sized factories, especially on the West coast, subjected to minimal or no environmental controls or requirements (Williams and Chang, 2008: 25; Rigger, 2011: 53). Like its neighbors, Japan and Korea, Taiwan has yet to solve the legacy brought about by the fast-paced development of the steel, paper pulp, cement and chemical industries in the 1970s decade (Tu, 2001: 230). This uncontrolled growth of small industries ended in the 1980s when the government shifted its economic strategies towards the development of industrial parks (Jobin, 2010: 46) or after business conglomerates began shifting their operations offshore (Tu, 2001: 255). Industrial parks had the adverse effect of intensifying environmental problems by leading to soil and water poisoning; a situation made worse by the fact that such parks were often located on fragile and unspoiled coastal lands, which were easily available, sparking debates in regards to land property issues (Williams and Chang, 2008: 135–165).

Under the political patronage of the United States Taiwan's economy boomed although the political environment remained oppressive and wary of dissent all throughout the 40 years in which martial law was in place (White Terror Era: 1947–1987) giving little space to voice opposition let alone petition on behalf of the environment. As mentioned, the majority of the people were not concerned about the damages and havoc brought upon the island by this sudden economic boom. Petrochemical plants and nuclear energy in particular came to be seen as a solace for Taiwan's thirst for natural resources (Hsueh *et al.*, 2001: 40). Taiwanese were satisfied with their improved living conditions and until the early 1980s the island enjoyed one of the highest per capita income rates among developing nations, accompanied by a vibrant cultural environment in music, arts, literature and film-making (Lupke, 2012: 157).

Alongside the changes at the economical level, a shift in Taiwanese's self-identification was gradually taking place, which had profound repercussions and reverberations on the cultural production sphere.

From the end of the 1960s and later on, in the 1970s, a humanistic trend, which emphasized the beauty of Taiwan's rural life, placing farmers at the heart of agricultural development as the symbol of a newly established Taiwanese Identity, started to appear in certain literary circles (Bain, 1993: 93). It became fashionable for writers and intellectuals to write about village life and "small" men and their plights as a valuable counterpart to city life and the "economic man." Lurking in the background was the implicit critique towards the KMT-led savage industrialization, considered as the culprit for having spoiled the natural environment and as morally "unfair" towards farmers, widely left behind in Taiwan's race towards economic development and modernity.

The Hometown Literature[1] movement fundamentally changed how Taiwanese saw themselves by awakening a strong sense of pride for their origins. As Shelley Rigger writes: "These revolutionary ideas brought a new willingness to practice a specifically Taiwanese way of life, to speak Hokkien without shame, to worship Mazu without apology, to eat the foods that flourished in Taiwan's rich volcanic soil without embarrassment" (Rigger, 2011: 33). These new ideas and strong cultural identification values are the same core components of the contemporary environmental movement. No matter whether a group is directly involved in helping small-town farmers in their struggle against multinationals such as Taiwan Rural Front, or whether it stresses the importance of supporting "natural" methods of cultivation used by Taiwanese farmers such as Homemakers United Foundation, the majority of Taiwanese green groups are deeply imbued by the need to protect the "native" cultural and physical richness of their country and its natural surroundings.

It goes without saying that those 30 years of unrestrained economic growth – without any checks and balances nor concerns for the environment – brought about disastrous consequences for the ecological well-being of Taiwan. As urban centers such as Taipei and Kaohsiung grew exponentially related problems such as motor vehicle traffic turned these cities into some of Asia's most polluted urban centers. Inadequate infrastructure such as wastewaters and sewage systems led to severe water pollution in rivers and water courses (Williams and Chang, 2008: 25) and the leap in population growth led to huge amounts of untreated solid wastes, which were amassed all over the island and inadequately handled, with catastrophic consequences for both human health and the environment. It was these first-hand and very visible effects that prompted mass action and fostered the birth of an environmental awareness.

The birth of an environmental awareness and of the environmental movement

It was not until the 1980s that a combination of circumstances led to the development of an environmental awareness in Taiwan. First, the visibility and scale of problems had become impossible to ignore; second, this was a decade of extreme mobility for Taiwanese who travelled abroad for leisure and even more importantly for working or studying purposes; overseas they obtained degrees that allowed them, upon their return, to connect the dots between economic development and environmental deterioration (Williams, 1992: 198). The growing environmental awareness among a minority elite of experts and scholars first and among the general population later on in the decade, prompted authorities to recognize the seriousness of environmental problems and the need to properly and urgently address such issues (Ho, 2006: 27). This process was fraught with problems, with several governmental bureaus experiencing conflicting interests that often pitted one faction against the other in fierce and prolonged battles. One such case took place over the Taiwan provincial Highway Bureau's plans to build another cross-island highway from the Western to the Eastern coast, which would have crossed one of last few wild areas left intact in Taiwan (Williams, 1992: 200). The Highway Bureau, in a classical developmental-faction alliance with various business communities and their representatives at the local level, was fiercely opposed by a group of conservationists allied with the Construction and Planning Administration and with the Department of National Parks. The project underwent five years of public debate and review and was finally suspended in 1986 but remains particularly important because it illustrates the growing influence of the environmental movement and the crucial role played by conservationists in those early stages, as the gradual and slow establishment of a process of open review and public discussion was difficult to come by.[2]

The predominant view among scholars of social movements is that the emergence of a plethora of new actors, which the state was reluctant to repress, starting from the 1980s challenged the long-standing authoritarian rule of the government and facilitated the creation of a public sphere, which these activists employed to exert pressure for the reorganization of the country's political structure (Wu, 1990).

All throughout the 1980s numerous social movements, addressing issues that ranged from women's rights, consumers' protection, aboriginal minorities' rights, environmental protection and labor issues, emerged. Beginning in 1988 free press and a variety of civic organizations and political parties blossomed. For environmentalists this meant being able to criticize the state for promoting economic development over ecological protection and demand better living standards (Ho, 2005a: 341). As noted by Hsiao, in comparison to other social movements, the environmental one was universal in nature and had a direct and strong impact on local politics (Hsiao, 1999: 41). Whether or not local residents would turn out to be successful in a confrontation with the local source of pollution depended on several external factors, as well as on their ability at mobilizing, by exploiting socio-cultural ties and networks; in fact the political majority, especially at the localities, was still firmly in the hands of the KMT. Therefore these early protests, rather than being aimed solely at the protection of a certain locality from polluting industries, were first and foremost attempts at gaining in-depth participation in the political life of the country, while trying to fight the KMT's previously undisputed domain over society and politics (Ho, 2006: 58–59). As a consequence, several such actions would quickly turn into anti-KMT movements, deemed culpable of favoring polluting industries over citizens, leading activists and residents to channel their sympathies towards oppositional parties. This gradual erosion of the one-party system slowly established the right conditions for the creation of a multi-party system, where different concerns and voices could be fully heard. As Hsiao puts it: "the environment-related anti-pollution, nature conservation, and anti-nuclear movements have contributed directly or indirectly to Taiwan's democratization, and this in turn has opened new space for these and other movements" (Hsiao, 1999: 43).

The increasing freedom enjoyed by the island after 1987 created a rapidly changing society in which new actors could suddenly voice their concerns. This is in fact the decade in which all of the most important grassroots green groups were born: *Taiwan Environmental Protection Union, Wild Bird Federation of the Republic of China, New Environment Foundation* and so on. These groups differentiated themselves from previous ones born out of local protests in self-interest because they were/are considered to be the first, advocating on behalf of the public (Tang, 2003: 1036). As stated by McBeath and Leng: "Three new patterns emerged during Taiwan's democratic reforms: legitimization of environmental protests, restructuring of political coalitions and decentralization, and incorporation of NGO participation in decision-making processes" (McBeath and Leng, 2006: 196).

As previously mentioned, this was also the time in which local officials acknowledged for the first time the seriousness of the island's environmental problems, pinpointing its causes in the uncontrolled growth of polluting industries as well as in citizens' lack of environmental education and awareness (Ho, 2006: 27). Governmental willingness to curb Taiwan's ecological distress led to the drafting of the island's first environmental policy plan, *Guidelines for*

Environmental Policies at the Current Stage, which was enacted in 1987, around the time when martial law was coming to an end and several struggles on behalf of the environment were becoming more vocal (e.g., the Lukang rebellion against a titanium dioxide plant of DuPont).[3] Struggles on behalf of the environment being less sensitive than direct and outward contention challenging the KMT's political rule, many dissidents chose to cluster under the umbrella-definition of "environmental activists."

Right after the abolition of martial law in 1987 social protests gained in frequency and violence until the early 1990s and then gradually decreased (Ho, 2006: 141; Jobin, 2010: 47). During the following years the KMT sought to regain control of society, and crack-downs towards the newly born opposition party (the DPP) and social movements were frequent (Ho, 2005b: 406; Linda Arrigo, interview, June 30, 2011),[4] increasing confrontations between the police and citizens and sparking several protests (Jacobs, B., 2012: 72). One such incident took place in 1991, as local residents and activists took to the streets in Gongliao to mark their opposition to the construction of controversial NPP-4, known as "*he si* 核四," inside Taiwan.[5] During the conflict one policeman lost his life, resulting in a national manhunt for what were termed "anti-nuclear criminals" (Linda Arrigo, informal communication, July 20, 2011). This radicalization of positions, either in favor or against nuclear energy, pressured numerous KMT members, actually opposed to nuclear energy, into silence for fear of being qualified as "DPP sympathizers." Not surprisingly, given the situation, green activists, and especially anti-nuclear ones, came to gravitate around the pan-green camp, initially perceived as a natural ally against the authoritarian and business-oriented KMT. However, this convergence of ideals between anti-KMT groups, which saw the DPP fully embracing the anti-nuclear cause, also symbolizes the beginning of what Taiwanese scholar Ho Ming-sho (2003) has termed "party dependence"; meaning that nuclear issues came to be seen as politically and ideologically identified with the green camp. While this initially helped activists find new outlets to disseminate their ideas, especially after the DPP won more than one-third of seats in the Legislative Yuan in 1992, in the long run it had the negative side effect of transforming nuclear issues into partisan issues, ascribed almost solely to the DPP (Ho, 2003).

Scholars of Taiwan's green movement generally agree that a true "environmental movement" began in earnest in the 1980s and specifically in the second half of the decade, after the DuPont incident (1986) and the lifting of martial law, when the expression of public dissent suddenly became possible. The victory of the small town of Lukang against a multinational American corporation in 1986 had an invigorating effect on Taiwan's infant environmentalism, which was fully realized after martial law was lifted one year later. As stated by Ho:

> Prior to the Lukang movement, Taiwan's environmentalism was mainly restricted to middle-class professionals who focused on educating the public and officials towards nature conservation's issues. Later on journalists and

> professors began working with the grassroots. The Taiwan Environmental Protection Union, established at the end of 1987, was the most representative product of this cross-class collaboration.
>
> (Ho, 2013: 701–703)

At any rate, by the end of the 1980s discontent towards governmental and industrial neglect and abuse was rising in numerous localities. This bottom-up civil force pressured the government into compliance, leading to the creation of a national environmental agency, the already-mentioned EPA, which was quickly followed by a complex hierarchy of agencies at all levels that dealt with various environmental issues. This gradually improved the situation, starting from the mid-1990s (Williams and Chang, 2008).

With the lifting of the martial law system and then Lee Teng-Hui's presidency in the 1990s, the democratization process was under way. After Taiwan implemented electoral competition and witnessed its first free and democratic direct presidential election in 1996 (Fell *et al.*, 2006: 15), the KMT was confronted with a double-edged sword; it had to sustain economic growth to win the elections while at the same time it needed to enforce stricter environmental rules in order to clean up the island. Indeed, throughout the whole 1990s environmentalists were empowered by the fact that democratization reduced the control capacity of the state sector.

According to Michael Hsin-Huang Hsiao's categorization of Taiwan's early movement, initially efforts were directed at three, partly overlapping issues of growing concern: nature conservation (since 1982), nuclear energy (since 1988) and environmental pollution (Hsiao, 1999: 32; Ho, 2005c: 238) derived mostly from excessive exploitation of natural resources from big enterprises (Tang and Tang, 1997: 281). In the 1990s collective actions towards local pollution became full-fledged movements in numerous communities, contaminated by various degrees of soil, air or water pollution. These movements started out as a reaction to an already manifested problem, either natural resources' degradation or health effects on people, by victims of this pollution and became later, thanks to media visibility, national grassroots movements. After the primary actors involved in such protests, namely the victims of pollution, a wide array of new participants from college professors and students to sympathetic intellectuals, soon joined the ranks of the nascent anti-industry and pro-environment movement. Opposition parties' members played a critical role in this period of early activism as well (Williams and Chang, 2008: 75).

The main difference between those early environmental protests and the contemporary ones, which will be analyzed in this book, is that the former were – for the overwhelming majority – *reactive*, with victims mobilizing to claim compensation for damages already suffered, rather than *pre-emptive*, as is the case for most protests in present-day Taiwan.[6] A crucial role in the organization of the opposition to polluting facilities, in those early days, was played by local temples; local deities, ghost festivals, folk goddesses and religious parades constituted an integral part of the action repertoire of local environmental mobilization

(Ho, 2005c). One of the notable examples of an anti-pollution protest, which stimulated a community response utilizing ritual and religious practices is the famous Houchin movement.[7] Nowadays the leadership capacity of maximizing its efforts in a concerted action, which transcends ethnic, gender, class and generational divisions has resulted in a more effective force for influencing governmental policies and pressuring authorities.

Another crucial difference between then and now is the target of protesters. In the early years of Taiwan's environmental activism complaints were mostly directed at big corporations and industrial parks with local residents uniting against the threat coming from the "outside";[8] local governments, deemed unresponsive towards environmental protection, were only secondary targets. Such a strategy changed quite quickly, starting from the early 1990s, as Taiwanese green activists have led projects, protests and watchdog activities playing a major role in shaping pollution control and conservation policies and frequently clashing with governmental authorities and industries. At the same time victims of pollution began asking for financial compensation and gave strength to their demands by employing radical techniques such as sit-ins to block entry to factories and such similar tactics (Ho, 2006: 62; 77). Business sector response was all but united, differing according to the size of the enterprise and the market covered, with the majority of companies aware that, in order to meet new stringent environmental regulations, production costs would have had to rise. Those enterprises which could afford to comply with governmental regulations (mostly big-sized ones) stayed in Taiwan and specialized in high-value items, requiring high levels of technological skills and know-how (Gereffi, 1990: 17) while small and medium-sized ones escaped the new requirements by transferring their production to cheaper countries, mostly in Southeast Asia and, a decade later, Mainland China. With this I am not suggesting that big-sized business conglomerates always comply with environmental regulations; rather, many of them still bet against the sincerity of governmental commitment towards the environment and make use of rising costs, pressure from green organizations and environmental pollution regulations as a blackmail tool towards the government, by threatening to leave Taiwan in search for cheaper countries in which environmental safeguarding regulations are not yet so strict (Hsiao, 1999: 44). The impact of such a bargaining strategy has been very high in Taiwan, a country in which the state was firmly convinced of the need of keeping big industries (e.g., the petrochemical and nuclear ones), as a sign of national sovereignty and strength to counterweight the waning international status of the country and the PRC's rising. Furthermore, the environmental movement also experienced a counter-movement by businesses, attempting to make environmental regulations less rigid (watering down the institute of the Environmental Impact Assessment, for instance) and/or launching large-scale counter campaigns against banning disposable bags and chopsticks, to name but a few of the most significant (Fell, 2012: 188).

The second current of Taiwan's early environmentalism, the "anti-nuclear movement" clustered under its banner countless political activists who sought to erode the KMT's four-decade long undisputed domain (Ho, 2006: 42). Their *cause célèbre* dealt/deals with opposition to the construction of a fourth nuclear power plant on Taiwanese soil, a project that dates back to the end of the 1980s, when the government first unveiled its intention to build such a facility, a story which is dealt with in Chapter 3.

The third stream in Taiwan's early environmental movement and also the earliest one to emerge, was the Nature Conservation Movement – henceforth NCM – (Hsiao, 1999: 36). Conservationists' main concern was to stop the damage to the island's precious natural resources. Affiliation and membership to the Nature Conservation Movement partly overlaps with that of the already expounded anti-pollution movement, since damage and havoc are brought on natural resources by industrial pollution and other human-induced activities; the difference though, lies in the appeal to higher moral values and the need to nurture nature and protect it from humans, which characterizes this third stream of environmentalism. The contrast with the anti-pollution movement, often considered as utilizing nature as an excuse to protect petty interests and individual properties, is striking. In fact, NCM's campaigns were aimed at saving endangered species and ecosystems such as migratory birds, mangrove forests, coastal wetlands, water and other natural resources from excessive exploitation and pollution. Key players, rather than local communities affected by the negative effects of pollution, were non-governmental organizations such as the *Wild Bird Federation of the Republic of China*, the *Homemakers United Foundation*, the *Green Consumers' Foundation*, the *Nature Conservation Union* and the *Society of Wildlife and Nature*, as well as numerous individual scholars. Women were an essential constituent of the nature conservation movement, keen on highlighting their role in society as educators and mothers, trying to lecture the general public on the need to nurture and respect nature, an important resource to be preserved for the future of our children (Chen Man-li, interview, July 20, 2011; Xu, 2011: 8–9).[9]

Furthermore the Nature Conservation Movement retains strong links with religious movements, in particular with several Buddhist organizations. The above-mentioned *Nature Conservation Union*, for instance, was established by a monk with the intent of saving Taiwan's coastal wetlands from the damages provoked by large-scale industrial pollution. Nowadays, an industrial belt of more than 100km in length renders most of Taiwan's West coast highly polluted for humans and animals as well as vegetation. A special role in promoting environmental awareness in a less confrontational manner was carried out by another association with a religious background, the *Buddhist Compassion Relief Tzuchi Foundation* (慈濟基金會).

Such organizations, with their efforts to solve environmental problems in non-confrontational ways, coupled with several changes affecting economical processes, such as the aforementioned emigration of polluting industries abroad and the ensuing industrial restructuring that followed, are all aspects that contributed to the slow but gradual establishment of an improved environmental situation. From the study of these early streams of Taiwan's environmentalism, we can observe how the anti-nuclear movement has been political right from the start (Ho, 2003, 2014a; Jobin, 2010) – therefore joining forces with oppositional parties such as the DPP – while the conservation movement has always striven to be perceived as non-politicized and neutral.

At any rate strict categorizations of the different streams composing the environmental movement have become outdated nowadays due to a partial – sometimes complete – overlapping between both leaders and members of many diverse groups (from anti-globalization forces to anti-nuclear ones), who have combined efforts in fighting for causes such as land evictions, polluting industries or super-imposed, controversial trade agreements. In the case of Taiwan's eighth petrochemical plant for instance, the anti-pollution and the conservation movements joined forces in view of the common goal of avoiding the destruction of a pristine wetland and its ecosystem and of pre-emptively fighting the forecasted negative health-damages for the local residential community.

The growing environmental awareness of the country is also reflected in the political platform of some of the more environmentally oriented parties, under Taiwan's "new," democratic, multi-party system, which has taken root in Taiwan since the watershed year of 1987. Among these, a prominent position is occupied by those parties belonging to the so-called Pan-green (翻錄 *fanlu*) coalition: the Democratic Progressive Party, the Taiwan Solidarity Union and the Green Party. First and foremost it is the DPP, formally established in 1986 by family members and lawyers of former political prisoners, that turned environmental protection into one of the core components of its political credo. However, as the party became more institutionalized, starting from the early to mid 1990s, its alliance with the environmental movement and its commitment vis-à-vis nature conservation gradually faded away, substituted by the need to boost the country's economy. More details in regards to the DPP's role towards the environment will be given in the course of the next section. The Taiwan Solidarity Union, established in 2001, claims to be in favor of a green Taiwan but is, according to several of my sources, putting developmental concerns first, just like the DPP (Lai Rong-xiao, interview, September 1, 2011; Lai Fen-lan, interview, October 30, 2011). Last but not least the Green Party, more on which will also be said in the course of the next section, established in 1996, enumerates among its members a small but fervent number of ideologically motivated adherents, who are active in the organization of protests, campaigns and vigils for several environmental issues.

Environmental issues under the DPP (2000–2008)

As previously stated, the link between politics and environmental issues in Taiwan should not be underestimated. Environmental concerns, among others, were an important driver for people to vote for the Democratic Progressive Party in the year 2000, thus ending 55 years of uninterrupted rule by the Nationalist Party. A mix of fortuitous elements, such as a strong earthquake that hit the country in 1999, revealing the low quality of buildings and cheap constructions all over Taiwan, and daily scandals of small-sized businesses disposing liquid and solid wastes into water bodies and landfills with impunity, ensured that environmental issues featured more prominently in the presidential elections campaign than in the past.

Even though with every election, pre-electoral canvasses and tactics will focus on warning one's electorate about the risks brought about in case of victory by the opposing party, in reality numerous issues tend to remain exactly the same, regardless of who is ruling the country. A good example is Taiwan's environmental deterioration, to which the DPP vowed to put an end, if ever elected, since at least the 1990s but continued unabated even after the party succeeded in winning its first presidential election in 2000. Before this watershed victory, Taiwan's future president, Chen Shui-bian, had promised to scrap controversial projects such as NPP-4, clean up polluted sites and enforce more stringent environmental regulations (Ho, 2005b: 412; Lai Fen-lan, interview, June 29 and November 24, 2011). Under direct pressure from either the DPP or from activists who had joined the party hoping to bring positive results, numerous provisions for environmental protection were indeed enacted. Until the mid-1990s, household garbage often festered in piles on public streets as trash collection was haphazard and there was no formal recycling policy.

Under the political patronage of the DPP, small cities and towns were cleaned up, a recycling system was established and increased transparency of operations and a more inclusive attitude towards activists became the norm (Arrigo and Puleston, 2006: 170). In fact, the DPP's earliest electoral campaigns were focused on delivering better quality of life at the localities by instituting modern transportation systems, cleaning up air and water courses, setting up parks and other recreational spaces and decluttering streets. The establishment of a capillary system for garbage collection, the setting up of numerous recycling facilities island-wide and the efforts to improve the ecological and residential environments of small towns and cities, drove many Taiwanese to vote for Chen Shui-bian, in the hope of replicating those positive local results at the national level. For environmental activists and eco-conscious individuals, the change in ruling parties represented a chance to redress the various imbalances that had led Taiwanese people and their rulers to favor economic growth over ecological concerns for over four decades.

However, after the DPP began its ascent to "institutionalization," scoring more than one-third of seats in the Legislative Yuan election of 1992, it gradually assumed a more cautious approach towards both social movements, wanting to change its image of an "unrefined" party (Ho, 2005b: 407), and towards environmental issues, considered as an obstacle to economic development. In fact, as a budding democracy with an industrial system in the making, the regime had to sacrifice its goals and compromise with big businesses in order to consolidate the country's economic development (Williams and Chang, 2008: 88; Jobin, 2010: 48, 52). Barely four years into Chen Shui-bian's presidency green activists were greatly disappointed with the ambivalence of many DPP legislators towards scrapping controversial plants or sacrificing economic goals to protect the environment (Lyons, 2009: 69). The rising unemployment rates and the worsening of the economy, coupled with corruption charges towards the administration, contributed to create a negative reaction towards the DPP presidency. In order to stabilize its precarious situation as "leading party" while the KMT continued to enjoy better local contacts and a capillary distribution of power at the localities, the DPP sought a compromise, which would make it more "acceptable" in the eyes of big business conglomerates, by sacrificing environmental protection (Arrigo and Puleston, 2006: 171). At the same time, mostly preoccupied with the country's economic setback, Taiwanese' support for environmental protection quickly diminished. As told to me by Paul Jobin, former director of the French Centre for Research on Contemporary China, CEFC, Taipei Office (2009–2013):

> Some of the most "liberalizing" actions towards industries were in fact met during Chen Shui-bian's Presidency; the legal framework concerning science parks was revised by the Legislative Yuan with the aim of releasing them from the control of local governments. As a consequence, these parks became grey areas, immune from laws such as the Environmental Impact Assessment Act.
>
> (Jobin, interview, October 4, 2011)

The DPP's change of attitude is best exemplified by former president Chen Shui-bian's behavior throughout the years. At his inaugural speech in 2000, his first announcements concerned his grandiose plan to turn Taiwan into a Green Silicon Island[10] (綠色矽島 *lüse xidao*) and the DPP's plan towards environmental protection, embodied in Taiwan's economic plan from 2002 to 2008, "Challenge 2008" (Williams and Chang, 2008);[11] barely one year later, at the Economic Development Advisory Conference held in August 2001, faced with the many financial and industrial difficulties of the country, he went so far as to claim that the EIA was a "roadblock to economic development" and that he "would like to 'kneel down' to the EIA reviewers on behalf of business" (as quoted in Ho, 2005a: 350). This Green Silicon Island slogan, chosen by Chen Shui-bian in the late 1990s, was meant to symbolize what the DPP strived to achieve if successful in the electoral competition of the year 2000 (Williams and Chang, 2008: 1). The basic idea behind it was the successful realization of Taiwan's industrial growth in information technology without damaging the surrounding environment. Nevertheless, as amply demonstrated by the objectives listed in the plan, its critics have vehemently argued that the DPP has betrayed its original commitment towards ecological safeguarding by focusing largely on economic goals, job creation and development of new infrastructures (Ho, 2005a: 342; Williams and Chang, 2008: 87; Tsui Shu-hsin, interview, March 24, 2011).

This change of attitude angered environmentalists who, in 1996, established a new political organ, the Green Party of Taiwan (GP), modeled along its German counterpart (Lai Fen-lan, interview, October 30, 2011), with the aim of capturing the sympathies of those who felt betrayed by the DPP's change of heart towards environmental protection. While the DPP was initially worried about this new political adversary, the GP failed to become a full-fledged political party, capable of threatening the DPP's electoral base (Ho, 2003: 703; Jobin,

2010: 48).[12] During the last presidential election held in January 2012, numerous Taiwanese had still never even heard of a "*Lǜ Dang*" (台灣綠黨 *Taiwan Lǜdang*). Even though the Green Party received almost 230,000 votes in January 2012, thus increasing its share nearly threefold – from 0.6 percent in 2008 to 1.74 percent in 2012 – making it the country's fifth political party, it still did not succeed in reaching the 5 percent threshold that is required to win a seat in the legislature (Chiu, 2012). The real "Green Party," in the minds of many Taiwanese, remains the DPP (Williams and Chang, 2008: 88). As aptly put by Paul Jobin:

> Despite its very limited number of members, the Green Party continued to play an active role in environmental issues, but more as a member of the environmental movement at large rather than as a political party *stricto sensu*, or as an alternative to the TEPU, which is too closely associated with the DPP.
>
> (Jobin, 2010: 48)

Even though numerous activists maintain that the DPP betrayed its promises, several positive results for the implementation of a green governance system were nevertheless reached under its aegis. As previously shown, while a first EIA law draft had been proposed already in 1983 it was a mere façade, designed to appease environmentalists' concerns while catering to the interests of businesses (Ho, 2004: 240). In 1994 a series of effective measures such as the acquisition of veto power on projects and increased public participation through public hearings, turned the EIA into an effective tool for individuals trying to access political decisions in regards to developmental projects. This was only made possible thanks to the changing political situation and the increasing number of social movements' leaders recruited by the DPP in the actual writing phase of environmental legislation drafts, thereby reaching a level of efficacy that was unprecedented. Other activists yet were assigned influential positions from which they were able to pass important pieces of legislation on behalf of the environment (Lyons, 2009: 62). But the most visible changes in the environmental realm took place at the local level, in townships and villages, where numerous DPP-dominated counties, more sympathetic and encompassing towards social activists, enacted new pollution-curbing regulations and recruited social activists into their ranks (Ho, 2005b: 407). However, with the increased institutionalization of social activists, formally enlisted in the DPP's ranks, environmental protesters shifted their *modus operandi* from the streets to more formal policy channels (Ho, 2005b: 410). More conventional "contentious politics" forms of protests have only recently been revived, in connection with the campaign on part of activists to stop the construction of NPP-4, which has seen a phase of resurgence since 2011, or against the island's eighth naphtha cracker (2008–2011), expounded in Chapter 3 and 4, respectively. Since the KMT's return to power in 2008, and especially after Ma succeeded in winning his second presidential election in 2012, many social activists and groups have resorted to more radical forms of protest and mobilization techniques of the past, feeling that their demands and complaints were being, increasingly often, ignored by the government (Cole, 2014).

Environmental concerns since 2008

In recent years the link between the environmental movement at large and the outcome of political elections and campaigns has become stronger. While both nuclear energy issues and chemical industry pollution had gradually lost importance in the 2000s (Jobin, 2010: 50), in light of the recent Fukushima nuclear catastrophe and planning efforts for the construction of a facility of the caliber of KPT (the already named eighth naphtha cracker), environmental concerns featured prominently during the pre-electoral campaign year in 2011 (Jacobs, A., 2012; Jobin, 2012). In fact, one of Tsai Ying-wen's (蔡英文) – the DPP presidential candidate in 2012 – basic tenets, consistently reiterated throughout her campaign, was the goal of turning Taiwan into a "nuclear free homeland" (非核家園 *feihe jiayuan*)[13] by the year 2025 (Tsai Ying-wen, 2011).

As Linda Arrigo and Gaia Puleston noted, one of the reasons why the KMT was unseated from power, after governing for more than 40 years, during the island's second democratic election was the mounting opposition against Taiwan's fourth nuclear power plant project, which the DPP had vowed to fight since the end of the 1990s (Arrigo and Puleston, 2006: 165). Afterwards, at the national level environmental issues were almost ignored both in the 2004 and 2008 campaigns, featuring more prominently in local elections; however, the numerous events of 2011 have brought back environmental concerns to the fore, turning them into crucial elements of political campaigning for politicians to attract votes.

Following a necessity to restructure in order to keep up with electoral competition, under Taiwan's newly established democratic system, the KMT has gradually transformed itself. Due to the changes in the international situation and the country's undefined status, it has gradually come to accept Taiwan as home, thus coming a long way from viewing the island as a short-term base and unreined-in industrial development as a *must* for its economic take-off. Nonetheless, certain clientele networks and attitudes typical of the authoritarian era, which accord priority to private interests and favoritism, are still a common feature of many "behind the curtain" deals taking place between developers and local governments, as we shall see with the four case studies. Such behaviors however, are no more an exclusive prerogative of the KMT, with the more radical fringe of the environmental movement and Green Party members considering both the DPP as well as the KMT as "part of the problem."

Starting from 2008 (Ho, 2010a) the environmental movement, as well as civil society in general, has seen a phase of resurgence and activism, especially visible among younger generations. The official incident, which marks the beginning of such a resurgence, is Chen Yun-lin's visit in November 2008 as the first official Chinese envoy to set foot in Taiwan in 60 years. His visit took place in a period in which numerous Taiwanese were beginning to think that their personal freedoms were slowly being curtailed by the state apparatus.[14] This return to semi-authoritarian and coercive forms of control sparked a wave of protest among students, who organized a big demonstration against police brutality; an event that was later portrayed in Chiang Wei-Hua's film *The Right Thing* and has become the object of study of several civil society scholars, who take it as the beginning of the resurgence of Taiwanese civil activism (Ho, 2010a). In general, as noted by Dafydd Fell, in comparison to the KMT of the 1990s, which chose dialogue over suppression and allowed for the gradual establishment of reforms that met the requests of civil society, such as direct presidential elections (Fell, 2012: 189), the latter administration has, so far, displayed a total disregard for and failure to respond to students' and civil society's requests on several issues, ranging from the abolition of the *Assembly and Parade Act* to demanding a clause-to-clause revision of the service trade agreement, thus creating the conditions for a more radical reaction on the part of the students, who felt that in no other way could their voices be heard.

As far as the environmental movement is concerned, starting from 2008 it has scored some groundbreaking victories. Among these is the success in obtaining – via a public referendum – the abandonment of a plan to have a casino built on the Penghu islands after the liberalization of the casino industry promoted by the KMT.[15] On an even bigger scale, between 2008 and 2011, environmentalists successfully opposed and stopped the construction of Taiwan's controversial eighth naphtha cracker, which was initially moved to Pengerang Malaysia (where China Petroleum Corporation hoped it would encounter less opposition) and finally canceled in 2013 by the developer for financial motivations. All the case studies analyzed in the book are perfect to illustrate this new-found vitality, with social movements frequently resorting to "old-style" methods like street protests, petitions, demonstrations in front of relevant governmental organs and concerts with famous artists as well as more advanced and novel participatory approaches that make use of available legal channels to protest inside the system. Despite the renewed vigor of protesters in fighting polluting complexes and the involvement of several high-profile lawyers who fight alongside with them, the case studies in this volume have brought to light the persistence of old clientele networks, which tend to "lighten" the burden of legal environmental obligations (such as the EIA process that developmental projects over a certain size have to pass) for big businesses, with the aim of providing them with cheap land and water supplies. Previous suppressive actions, employed against those who opposed naphtha crackers, industrial parks and nuclear plants guilty of destroying wetlands and fertile lands around the island, have gradually been substituted by a war of experts with contrasting opinions, on the payroll of each faction.

Finally, one of the biggest changes that has taken place in the past few years in the environmental sphere, is the growing importance accorded to ecological issues as topics of national prominence; a focus that has grown stronger after the watershed year of 2011 with both parties competing to include green issues in their presidential platforms for the 2012 elections.

Before diving into the analysis of the actual case studies it is necessary to get better acquainted with one of the key players in this ongoing environmental struggle with the state: namely green activists, through the information gathered via several in loco interviews and analysis of relevant literary sources.

Environmental activists in contemporary Taiwan: achievements and difficulties

As previously shown, in the early days Taiwan's environmentalism was comprised of different actors: from middle-class intellectuals and scholars educated abroad – aware of the need to protect the environment and in tune with the international environmental community and the dominant environmental discourses – to victims of pollution, whose scope and interests rarely extended beyond that of their immediate surroundings (Ho, 2012: 6). These two groups sometimes collaborated and sometimes came into opposition as they were, to borrow from Ho (2012: 6) "operating in [a] different universe of meanings." Frictions were particularly severe when monetary compensation for pollution, viewed by middle-class intellectuals as a "selling-out" of the environment, was involved. While such tensions were extremely high in the 1980s and 1990s, with mutual incomprehension fostering suspicion and pitching one group against the other (Ho, 2006: 52), the situation has, since then, gradually improved and there is currently abundant evidence of cooperation among the different groups comprising the environmental movement, facilitating a positive outcome for protesters. Furthermore, a plethora of newer figures, from the lawyer/activist to the members of a more radical anti-globalization fringe have recently joined the ranks of Taiwan's environmentalism.

In the various cases analyzed in the book, especially those that take place in rural areas, where local residents are less informed in comparison to their urban counterparts, middle-class professionals and social activists have cooperated closely with rural villagers, to educate them and provide them with legal aid and assistance. The "elitist" attitude of environmental activists, more preoccupied with providing rich urbanites with an improved urban environment than dealing with the plights of the rural population, has been radically transformed by this new generation of green activists, who target rural villagers as main recipients of their aid efforts; most notably, in this regard, is the already named Taiwan Rural Front.[16] Rural residents, on the other hand, have come to realize that their chances of victory against powerful business lobbies are greatly diminished if they fight alone, thus they eagerly welcome the assistance of better-established urban activists and NGOs. While the environmental movement has come a long way in Taiwan, becoming stronger and unified against powerful industries, vicious factionalism can still be found among green activists, often proving detrimental to the promoted cause; such was the case in the Taipei Dome developmental project, which will be illustrated in Chapter 5, where a fruitful cooperation between different groups was partly hindered by the quarrels of the main protest leader with one Green Party member and with a committed lawyer, who had offered his help pro bono. This capacity of making environmental concerns "trans-local" by establishing broader coalitions, able to garner nation-wide attention, has simultaneously rendered the environmental movement less partisan and more autonomous as well as establishing the conditions for a mounting national support towards their campaigns. Particularly significant is the skilled employment of key experts on the part of activists, such as lawyers fighting through the employment of rules and regulations, and medical experts and epidemiologists, releasing several studies on past and future (forecasted) effects of pollution on human health, which enhance their battles' credibility.

While during the authoritarian era it was the KMT that defined activists as "hooligans," many of my sources in the environmental movement have used the word *liumang* (流氓) to describe the intimidating and rogue techniques employed by both construction companies and local governments to make them desist in their quest to stop construction projects. Some of my interviewees claimed to be afraid of physical attacks and of threats still being common (Chen Man-li, interview, July 20, 2011). For instance, Chen Chiao-hua (陳椒華), long-time environmental activist, was attacked in 2011 by masked men in Tainan for her participation in a campaign against a local garbage dump in Dongshan (東山) District (Chen Chiao-hua, interview, July 28, 2011). The incident sparked a wave of protests on part of environmentalists all over Taiwan. Another source, who prefers to remain anonymous, related several silent phone calls she received at night, when protesting against a high-profile controversial project involving relevant sums of money, and numerous instances in which her car tires were slit (anonymous informant, interview, March 14, 2011).

According to the *Taipei Times* and to some of my sources, several environmental activists – including *Wild at Heart Legal Defence Foundation* founder Robin Winkler (文魯彬 Wen Lubin), the foundation's secretary-general Lin Tzu-lin (林子凌), Green Party of Taiwan's former spokesman Pan Han-sheng (潘翰聲) and TEPU former president, Wang Chun-shou (王俊秀) – also claimed to have been either physically attacked or verbally threatened on several occasions (Loa, 2011; Pan Han-sheng, interview, September 1, 2011; Robin Winkler, interview, September 5, 2011).

In spring 2013, a couple of incidents where environmentalists trying to protect trees from being transplanted or felled were manhandled and handcuffed (with policemen seemingly misplacing the handcuffs' keys) have sparked concerns in regards to a possible return to semi-authoritarian forms of ruling and police brutality (Shih, Y.H., 2013). One positive note comes from Chen Man-li, the already mentioned chairwoman of the *Homemakers United Foundation* who, during our conversation, remarked how, in comparison to the 1990s when for activists it was quite dangerous to be working in the field of environmental protection, the situation has nowadays greatly improved, even though they still occasionally receive silent phone calls or threats, especially during a sensitive campaign (e.g., nuclear power opposition) (Chen Man-li, interview, July 20, 2011).

Nevertheless certain "performance" problems remain; while it is undeniable that Taiwan has steadily improved its environmental conditions, "green projects" are often showcases for boosting either the state or corporations' images. Miss

Chen shared with me her concerns that to this day, the real figures in regards to the collection of recycled garbage are actually being forged (Chen Man-li, interview, July 20, 2011). Her claim struck a chord when, a few weeks later, I was sitting under a shady tree in the lush green park of National Chengchi University (政治大學 *zhengzhi daxue*) with a friend, trying to find some solace from the scorching heat of a torrid August day. While we were lazily chatting away a garbage truck came by ready to collect recycled goods. In universities, as well as in many other public places in Taiwan's urban centers, citizens are required to recycle items that range from plastic bottles to glass and paper products. Schools enforce strict regulations and the same degree of severity is applied to metro stations and other public areas, with violators being subjected to heavy fines and penalties for infringing the law. Adding insult to injury is the overall lack of trash cans in Taipei, which puts eco-friendly citizens in front of a choice: either renounce buying something to drink – easier said than done given the abundance of 7/11 and convenience stores virtually everywhere – or end up holding their styrofoam cup for as long as 50 minutes without being able to discard it! To return to what I witnessed on that hot summer day, while the initial process is flawless, with students and staff collecting their items and neatly stashing them in differently labeled containers, my friend and I witnessed how the garbage truck, after picking up the trash, regrouped all items in a common pool without any distinction being made. Questions arise as to whether these trash items are later to be divided once again (most probably not) and why then do students and teachers have to go to such great lengths to divide their trash, only to have someone regroup all the items again! A possible explanation I received from Chen Man-li pointed to the government wanting to boost its image of an eco-friendly administration, with a responsible attitude towards basic fundamentals of environmental protection, such as recycling activities, without any real commitment (Chen Man-li, interview, September 21, 2011).

These, as well as other previously mentioned problems, still pose an obstacle to a proper functioning of an environmental protection system worthy of a modern and wealthy country such as Taiwan. While the country has come a long way in solving some its most abysmal environmental problems, brought about by 30 years of uncontrolled economic development, politics have become so intertwined in environmental processes that it is often unclear which parties are truly committed to the environmental cause and which are exploiting and heralding green issues simply to attract more votes. This aspect will be further expounded through the analysis of four emblematic case studies in the course of the following chapters; suffice to say that while everywhere in the world the political system is closely intertwined with economic transformations (and thus environmental damages) in Taiwan this dimension is even stronger. Paradoxically though, even though environmental issues have been playing a fundamental role in Taiwanese politics for the past 30 years (in the process of institutionalizing democracy, in the DPP's electoral strategy – and lately also in the KMT's), in practice, though often sugarcoated in eco-friendly and green discourses, the dominant imperative still seems to be "business as usual."

A positive turn of events is to be found in the recently occurred emancipation of the environmental movement from the DPP, after several years of close collaboration, which hindered environmental and ecological issues from becoming cross-partisan. Partly because of anti-nuclear concerns and partly owing to activists' disenchantment with the DPP's true stance in regards to NPP-4 and environmental issues, several younger activists have distanced themselves from their forerunners, putting an end to the idea that ecological safeguarding is an exclusive prerogative of the DPP and have grown nearer to the Green Party and other smaller and more independent organizations.

Conclusions

While the early democratization process of the 1980s legitimized environmental protests for the first time in Taiwan's history, incorporating public participation and requests in the policymaking process, the DPP's first electoral term propelled the country back towards a business-oriented mentality. The party, early on the most natural ally for environmental activists, has meanwhile come to be viewed with suspicion on the part of environmental activists, as guilty of having "betrayed" its original commitment.

Today Taiwan is politically a deeply divided country; the two main parties and their allies all share different views on a variety of issues ranging from economic development and environmental protection to the island's future, which render governmental efficiency volatile and the electorate fragmented; this, in turn, makes it more difficult for environmental agencies to implement regulations that could effectively protect the environment. Nevertheless, Taiwan has come a long way in its economic development and has borne the brunt of its initial unrestrained economic growth. For several decades decisions were taken from the top down without seeking any consultation among citizens thus producing an unprecedented damage to land, soil, air and water resources not to mention human health. However, in parallel with its impressive race towards democratization, increasing ecological and human rights' awareness has gradually but relentlessly driven citizens to expect a higher quality of life and to demand respect for the surrounding ecosystem. Three decades of environmental activism have transformed and rooted Taiwan's environmental consciousness. Cities are less polluted and more livable, a system of public transportation has been established and several polluting facilities have relocated away from Taiwan.

Increased wealth and democratization have brought about environmental improvements, but many aspects that would ensure an effective and efficient system of ecological governance, such as low levels of corruption and a solid regulatory system, are still either missing or hampered by hard-to-die bad habits, "endemic" to Taiwan. The following chapter will focus on one of the most controversial projects inside Taiwan, which has sparked conflict and problems in an endless battle that has been raging for more than a quarter of a century: Taiwan's fourth nuclear power plant.

Notes

1 *Xiangtu Wenxue* (鄉土文學), literally "Literature of the Village Soil," despite the multi-faceted and interconnected meanings it has in Chinese, is used to define the literary genre that focuses on the agricultural, rural and traditional way of life rather than on urban, modern life.

2 As shown by Ho in his comprehensive study of the environmental movement in Taiwan "Green Democracy" (綠色民主 *Lüse minzhu*), conservationists were crucial in helping to establish a different cultural mind-set and for changing what he terms as "Chinese bad habits," meaning, for example, the culinary custom of eating wild animals and migratory birds, to this day still considered as "delicacies" in the Chinese cuisine. In 1979, precisely the protection of migratory birds sparked one of the earliest confrontations between conservationists and locals in Kending, Hengchun peninsula. The designation of the Kending area as a "protected national park" in that same year had brought on severe limitations to locals' consumption habits (Ho, 2006: 39).

3 This incident refers to a protest against DuPont's plan of building a petrochemical plant in Lukang (鹿港), which ignited opposition by local residents and officials that resulted in DuPont's voluntary withdrawal of the project (Reardon-Anderson, 1997).

4 Linda Arrigo is an historical figure in Taiwan's struggle for democracy and human rights with deep ties to the environmental movement. In the past she acted as the international spokesperson for the Green Party of Taiwan. She is currently working at Taipei Medical University as an associate professor.

5 In 2004, Tsui Shu-hsin, secretary-general of GCAA, produced a documentary on the lives of Gongliao locals, shattered by the incident that had taken place in 1991. The acclaimed documentary is called "Gongliao, How Are You"? (貢寮你好嗎? *Gongliao ni hao ma*?).

6 According to Hsiao (1999: 35) in the 1990s the percentage of pre-emptive protests was at about 16 percent, nevertheless a significant increase in comparison to previous decades in which no protest of the pre-emptive kind took place, indicating a lack of trust towards the government's willingness to solve the island's abysmal environmental situation.

7 Houchin (後勁) is a village located in the vicinity of Kaohsiung, where local residents opposed an expansion project by China Petroleum Corporation (CPC) in 1987 and started a protest that lasted three years (Weller, 1999: 115–120; Ho, 2005c: 239); elements of folk religion made up a significant portion in the collective action repertoire employed by protesters.

8 This being part of the reason why Taiwan's local anti-pollution movement has been defined as being a "populist" type of movement (Hsiao quoting Szasz, 1999: 35) in which actions are usually started at the local level – with little initial help from outsiders, such as green organizations or concerned intellectuals.

9 Chen Man-li, chairwoman of the *Homemakers United Foundation*, told me during our first interview (on July 20, 2011) that the initial spirit of the foundation, as it was established in 1987, was to show that women could be something more than mere housewives, caretakers and cooks; they, too, have an important role to play in society *outside* of the family. Again, as is often the case with pioneering environmental movements, rather than being solely aimed at protecting nature, such ideas and concepts were first and foremost attempts to change the predominant mentality of those times, which considered women's duty and goal in life to be dutiful spouses for their husbands and good mothers for their children. Thus, such associations played a crucial role in trying to raise women's status in society from de facto "second-class citizens" to a position of parity with that of the other sex (Chang, 2009).

10 A reference to California's Silicon Valley, based on the fact that Taiwan has built its fortune on the semiconductor and electronic microchip industries.

11 For information on the Six-Year National Development Plan see Shapiro (2002).

12 Its core membership is varied with a good selection of affiliates representing aborigines, reform activists and environmental activists.

13 It is my impression, gathered by talking to several people from all walks of life during the year 2011, that such claims on part of Tsai were perceived by many as naïve, not supported by facts and lacking any basic explanation as to what kind of energetic sources Tsai would use to substitute nuclear energy, which prompted many to think of such declarations as simple electoral campaigning strategies, devoid of any real commitment.

14 A notable example is to be found in Professor Ho Ming-sho's discomfort in being contacted by the Environmental Protection Administration, expressing the agency's concerns over an op-ed article he had written (Ho, 2010a: 17; Ho, interview, August 30, 2011).

15 In July 2012, a similar referendum was held on Matsu Island with 54 percent of local residents voting in favor of the central and local governments' plan to build a casino in even closer proximity to the mainland, in order to boost Matsu as a gambling and tourist destination (Our Island, 2012a). The casino is scheduled to be open at the earliest by 2019 (Chen, 2013: 89).

16 It has to be noted that it is increasingly frequent for young, charismatic protest leaders to travel around Taiwan to explain their group's strategies while trying to garner support for their cause among the rural population. Particularly active in this regard have been the members of the Sunflower Student Movement.

3 Nuclear energy and the fourth nuclear power plant controversy

A short story of nuclear energy in Taiwan and the birth of the anti-nuclear power movement

Nuclear power seemed to be a promising form of energy in the aftermath of World War II, because it guaranteed clean and cheap energy supplies, which would not wreak havoc on air and water, while helping power-hungry nations recover from the disasters of the war. In 1955 the Taiwanese government established the Atomic Energy Council – AEC – (原能會 *yuanneng hui*), which would become the leading organ in nuclear-related issues in Taiwan, responsible for radioactive medical wastes, nuclear wastes' safety regulation and radiation protection among other tasks. Back then nuclear energy was seen as a cure-all solution to the growing insecurity surrounding oil and was viewed as a matter of national sovereignty, which had to be included in the so-called "Ten Major Construction Projects," a state-led program to industrialize the country (Ho, 2014a: 4).[1]

Taiwan entered the nuclear energy business during the KMT era, a period during which decisions considered crucial for the nation's economic well-being were taken by the government without prior discussion, nor any attempt to search for consent among citizens; at that time priority rested with industrialization, which in turn demanded a large amount of cheap energy (Williams and Chang, 2008: 72). On this last point there is speculation among experts and environmentalists that the drive towards nuclear energy was not entirely domestically dictated but at least partly externally driven by countries such as the United States and England, which had a strong interest in finding new markets for the sale of their nuclear reactors; Taiwan was one of these countries (Hoare, 2013: 29).[2]

For all of these reasons, especially in the 1980s, a decade in which the state placed a strong emphasis on economic development, nuclear power generation came to be seen as indispensable to sustain high growth (Hsiao, 1999: 38). By 1985 Taipower had already finished building all three existing nuclear power plants and was planning a total of 20 reactors on the island by the year 2000. The total number of currently functioning reactors in Taiwan has stopped at 6.

Construction of the first plant began in 1971 in Shimen (石門),[3] Taipei County (which changed its name to New Taipei City in 2010), the second plant

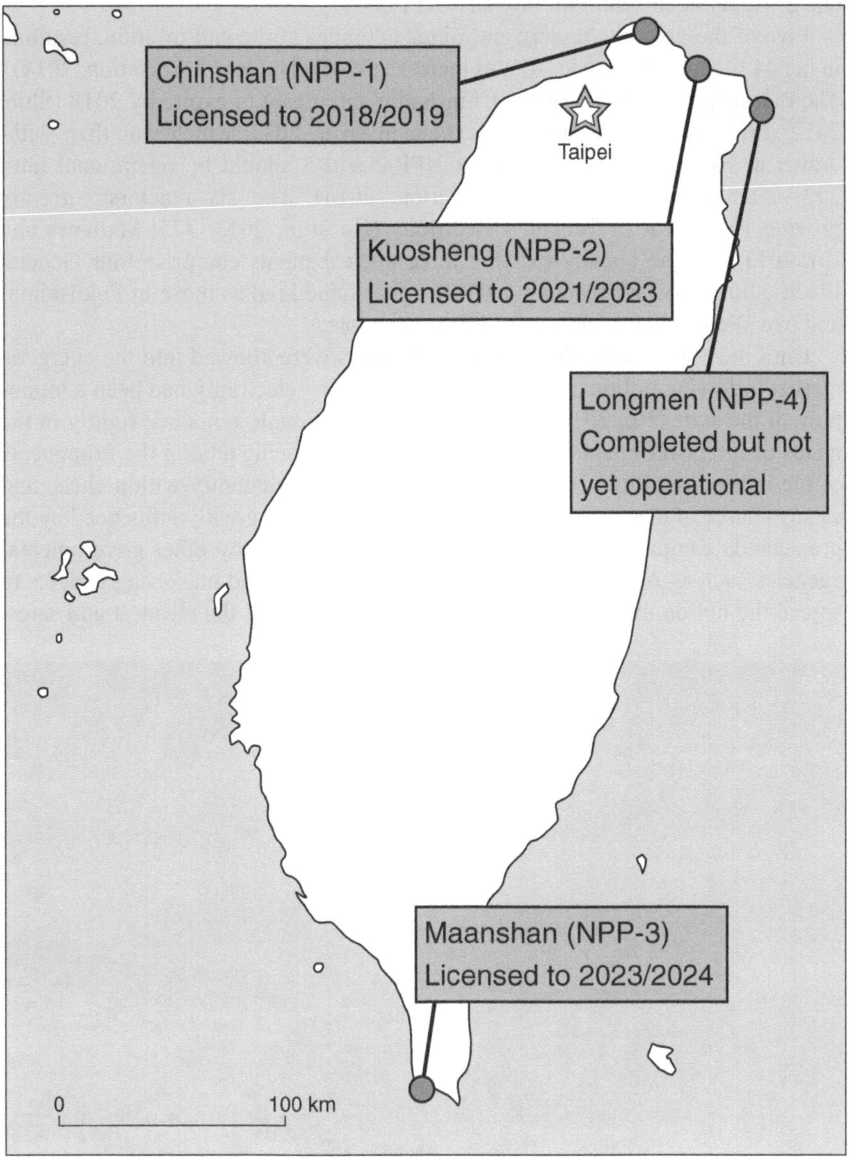

Figure 3.1 Geographical location of Taiwan's power plants (source: Simona Grano).

started in 1974 in Wanli (萬里), also in Taipei County, while construction of NPP-3, set in Hengchun (恆春), Kenting (墾丁), began in 1978. The three plants are all operated by state-owned Taipower – under the Ministry of Economic Affairs – and are expected to have a 40-year lifespan. They have been operational since 1978 (NPP-1), 1981 (NPP-2) and 1984 (NPP-3) (Ho *et al.*, 2013:

775). Geographically, two power plants are located in northern Taiwan, NPP-1 and 2 – and one in southern Taiwan – NPP-3.

Five of the six units underwent minor upgrades at the end of 2008, resulting in net 44 Megawatt electric (MWe) increase (World Nuclear Association, 2014).[4] The oldest reactors, those of the Chinshan plant, meant to expire by 2018 (Shih, 2012: 296), had been granted an extension up to 2037, which was then withdrawn in November 2011; those of NPP-2 and 3 should be operational until 2021–2025 (World Nuclear Association, 2014). The six reactors currently provide 19 percent of Taiwan's electricity (Ho *et al.* 2013: 775; Mathews and Hu, 2013: 2). The already existing three nuclear plants comprise four General Electric boiling-water reactors – BWR – (the same kind as those in Fukushima) and two Westinghouse pressurized water reactors.[5]

Until the 1990s, when independent providers were allowed into the energetic market following political liberalization, generating electricity had been a monopoly of the state (Ho, 2014a: 4). Nuclear energy though, remained tightly in the hands of the government, with Taipower and AEC being among the proponents of the indispensability of nuclear power to ensure the country with a cheap and steady source of electricity. Public opinion was initially greatly influenced by the propaganda campaigns financed by both Taipower and by other governmental agencies such as AEC, which produced books, comics and marketing gadgets to spread the notion that nuclear power stations represented the cleanest and safest

Figure 3.2 NPP-3, located right behind a famous beach in Kenting's national park, August 11, 2011 (source: Simona Grano).

form of energy and that radiations had several useful secondary effects, such as for instance making fruits and vegetables grow bigger (Schafferer, 2001: 103). Even though such propaganda efforts were initially successful, this mind-set changed quite quickly when problems such as how to deal with nuclear wastes and incidents such as Three Mile Island in the United States (1979) and Chernobyl in Ukraine (1986) made nuclear issues (and related fears) known to the general population, stimulating opposition. Furthermore those were the years in which several Taiwanese scholars and students who had left for the United States to pursue their degrees, finally returned home. Some of these American-trained experts and scholars began questioning the KMT-led nuclear expansion plan and started to spread the global anti-nuclear discourse in Taiwan (Weller, 2006). Growing support for the anti-nuclear camp was further helped by a series of incidents and stalls that took place in the early 1980s. In fact it was a fortuitous fire that had taken place at NPP-3 (whose budget ended up costing double the price of the originally approved sum) in 1985, which together with the Chernobyl incident, delivered the first serious blow to the government's nuclear expansion plans (Schafferer, 2001: 99). A few months later, during a debate organized by the *Homemakers United Foundation*, one of the island's earliest organizations, Taipower officials were subjected to extensive quizzing by intellectuals and activists, worried about the feasibility and safety of the newly proposed plant, namely NPP-4.

While at first anti-nuclear voices came almost predominantly from academics and intellectuals after 1988, benefiting from the increase of individual freedoms following the lifting of martial law, local residents of Gongliao joined the protest, de facto symbolizing the birth of the Taiwanese anti-nuclear movement, which came to comprise concerned local residents, intellectuals and civil society organizations (Hsiao, 1999: 38; Ho, 2006: 74). Due to Taiwan's history of democratization, once martial law was abolished, creating numerous outlets to vent discontent (newspapers, television programs, public protests), many of the previously repressed social movements and opponents chose the anti-nuclear cause as their *cause célébre* to attack the state.

One of the most active players in the early stages of the anti-nuclear movement was the afore-mentioned TEPU, established by a group of university professors and students, mostly congregating around Taiwan National University,[6] in 1987 (Shih, 2012: 301). While TEPU's commitment to the anti-nuclear cause is still strong up until today, in the past decade newer organizations, such as GCAA and *No Nuke Cultural Activism Group* have enjoyed greater visibility and play the leading role in the organization of the several anti-nuclear protests, flash-mobs and online events that took place between 2011 and 2014, as we shall see in the following sections.[7]

Public awareness of nuclear energy and related risks in Taiwan

The degree of public acceptance of nuclear power differs from country to country and depends on the historical background of each place. Therefore, in some countries citizens are more passive and graciously accept nuclear plants (e.g., in France where the government has successfully spread the idea that it is necessary for the economic well-being of the country, and in Switzerland or Finland, where trust in governmental authorities is high) while in other places (Italy and especially Germany with its long-standing "green" tradition) citizens are active and proactive in battling new or even already existing plants. While the Fukushima Dai-ichi incident in 2011 sparked numerous reactions both in the Western world and in Asia, it is in the latter that several governments (e.g., the PRC) have decided to forge ahead with their nuclear expansion plans. Taiwan, whose geographical position and seismic nature are similar to those of Japan, has witnessed some changes in state-led energetic policies and decisions in the past few years, which are the direct outcome of certain outside events, such as the Fukushima incident, that will be analyzed in depth towards the end of this chapter.

As mentioned, different countries experience different historical evolutions, which have profound reverberations on how its citizens perceive nuclear energy. To fully understand the situation in contemporary Taiwan, a few elements, pertaining to the formation of a nuclear awareness in the ROC, have to be discussed first.

First of all, how did residents of the localities where power plants have been built, react to such "siting" decisions? As explained to me by Ho Ming-sho, when discussing the first nuclear power facility set up in Taiwan, local residents in Shimen initially welcomed NPP-1 and the new opportunities (in terms of job creation) it would bring (Ho, interview, August 30, 2011). However, barely one year after the plant had become operational in 1978, attitudes changed after villagers discovered that the huge heat release from the ocean water cooling system, coupled with the powerful chemical products employed to keep the pipes clean, had significantly diminished their catch of fish. A related issue, which contributed to add insult to injury for local residents, was the imposition of a ban on any new development in the area within five kilometers of the plant, which meant that locals would be excluded from the financial benefits derived from the rising prices of the real estate market, in areas located just outside the capital, such as Shimen, Chinshan and Wanli (Ho, 2006: 74).

A similar situation took place in the late 1980s and 1990s in Gongliao, after this location was selected for NPP-4. Since locals were already aware of the potential damages that such a facility could create, due to the previous experiences of other rural communities, such as Shimen and Wanli, a fierce protest had been mounting since at least 1987 as intellectuals, *Dangwai* (黨外) members and anti-nuclear scholars such as Chang Kuo-lung and Lin Jun-yi started to hold anti-nuclear speeches, informing the local community about the risks that awaited them and providing information of various kinds (Ho, 2003: 694).

In the early 1990s Taiwanese people's awareness of nuclear risks was further heightened by a long-kept-silent scandal, involving nuclear steel rods. From 1982 to 1984 radioactive scrap metal and steel, scavenged from the nuclear plants, were melted into rebar and used in the construction of apartment buildings and schools in several counties in Taiwan.[8] The contaminated steel was purchased by a company named Shin-long, which smelt it and re-sold it to various steel-processing industries in northern Taiwan. Over 2,000 apartment units and shops are thought to have been built with such contaminated materials and at least 10,000 people are known to have been exposed to long-term low-level irradiation as a result (Chiu, 2001). More precisely, experts estimate that 15,000 citizens and students have received excessive low-dose-rate of gamma irradiation for between two and 15 years (Chang *et al.*, 2001: 655).

This situation was not publicly known until the year 1992. High amounts of radiation had actually been detected as far back as 1985, when a dentist set up an x-ray machine in his apartment in the Minsheng Villas (民生別墅 *minsheng bieshu*) in downtown Taipei. When the AEC went by to conduct a radiation-safety surveillance check, it detected dangerous amounts of radiation even though the x-ray machine was not plugged in (Arrigo and Puleston, 2006: 177).

The two officers dispatched to carry out the safety check on the machine submitted their reports to the AEC on April 1, 1985. AEC then charged two specialists with the task of tracing the source of the construction steel (Schafferer, 2001: 104). All results of the investigation were reported to the then director of the AEC, who decided to take no action. Rather than telling the truth – that the radiation came from the walls – AEC banned the dentist from operating the x-ray machine. The truth came out several years later, in 1992, thanks to a reportage by the *Liberty Times* (自由時報 *ziyou shibao*). Only then did Minsheng villa's inhabitants first discover that their properties were radioactive. This fame sadly haunts this place today, with the Mingsheng villas having meanwhile been renamed "radioactive houses" (輻射屋 *fushe wu*) (Our Island, 2012b).

In an effort to help the victims seek redress for their grievances, Professor Peter Wushou Chang (張武修) led a research team at National Yang Ming University (陽明大學 *yangming daxue*) for five years (1995–2000), which kept track of more than 4,100 individuals who once lived in buildings that had been constructed in Taipei City between 1982 and 1984, using radioactive steel reinforcing bars (Peter Wushou Chang, informal communication, August 31, 2011). His study, concentrating on the incidence of diseases among residents of radiation-contaminated buildings, showed that the possibility of chromosome aberration (damage to DNA) was proportional to the long-term exposure of the subjects to low-dose radiation. However, even though the team observed a high incidence of cancer, especially cervical cancer, breast cancer, liver cancer, leukemia and thyroid cancer, in the subjects analyzed, due to the length that such studies require to turn out to be correct, it was ultimately impossible to conclude that there was a direct link between the cancer rate and such patients' exposure to low-dose radiation (Chiu, 2001: 2).

The radioactive houses' story, sadly famous inside Taiwan, has contributed to spread knowledge and fear of the potential risks of an industry that had been, for a long time, similar to the petrochemical one, considered as "bringing prestige" to the country and as being "closely linked to the issue of sovereignty." Further, this episode serves as a tragic memento of the non-transparent behavior of the nuclear regulatory commission and governmental authorities not too long ago and of how such actions have tragically destroyed the lives of innocent citizens.

NPP-4: a political controversy

The anti-nuclear debate and the long-standing fourth nuclear power facility project best exemplify the degree of continuity characterizing numerous issues in the environmental sphere, in spite of changes taking place at the political level. This controversial plant, by now built in its entirety, has encountered public opposition and a host of delays since its very first planning phase. In fact, while Taiwan's first three nuclear power plants were all built before the lifting of martial law, and thus with no involvement nor opposition from the public, the KMT government first announced its intention of building a fourth nuclear power plant in the 1980s, as the authoritarian one-party rule was slowly relenting its grip on society (Lupke, 2012: 159–160). Actual construction of the facility started in 1998 but the project has been marred by political controversies and stalling. If it ever becomes operational, NPP-4 will feature two more reactors, bringing Taiwan's total capacity to eight, belonging to the Generation III advanced BWR technology. While the plant was supposed to become operational by 2015 (United Press International, 2013), after the Sunflower Student Movement and the Lin Yi-hsiung strike in the spring of 2014 – two episodes on which more will be said at the end of this chapter – construction of the facility has been put on hold until a popular referendum will most probably settle its fate.

The saga of NPP-4 is a never-ending story of intrigue, political corruption, legal violations and drama that has outlasted two changes of ruling parties and clearly illustrates the strong link between environmental and political issues in Taiwan.

NPP-4 is also the emblem of the anti-nuclear movement's struggle in the country. Since the KMT government first announced its intention of building NPP-4, the project has been nearly canceled twice; people involved in its construction and planning have ended up in jail following corruption accusations and at least two persons have been sentenced to death because of it (Williams and Chang, 2008: 72; Lupke, 2012: 159–160). Its saga is widely known among the general public since its story, depicted in detail by the media, has been dissected virtually from every possible angle, from journalistic reportages on television show Our Island (2011c) to protests taking the form of politically dissenting documentaries such as already mentioned "Gongliao, How Are You?" by Tsui Shu-hsin, secretary-general of *Green Citizens Action Alliance*, the previously mentioned advocacy-focused grassroots movement, which deals with anti-nuclear issues, among other things.

Figure 3.3 Tsui Shu-hsin and Wang Chung-ming during a protest for the protection of Tamsui's mangroves, September 18, 2011, Tamsui (source: Simona Grano).

Since its very first planning phase, as political relaxation and democracy were settling in (after 1987) protests clustering around the Gongliao plant came to be characterized by a unique mix of elements from political power, public involvement, local and central government frictions and grassroots activism that would have been repressed in Taiwan only a few decades earlier (Lupke, 2012: 157). According to Michael Hsin-huang Hsiao, to justify the need for a fourth facility, the KMT sought to manipulate public ignorance in regards to nuclear energy by instilling the belief that due to energetic shortages, a fourth nuclear plant was inevitable. Hsiao also claims that Taipower has duly complied with this version of events by rationing energy whenever the anti-nuclear movement appeared most influential among the public (Hsiao, 1999: 49). However, more needs to be said about the historical circumstances surrounding the choice of Gongliao and the early planning phase for this facility.

Taipower had purchased land in Gongliao in 1983. Following the international reverberations of nuclear incidents such as the one in Pennsylvania, a group of activists and intellectuals started to fiercely criticize the government's intention to build a fourth nuclear power plant (Ho, 2006: 43, 74). In 1985, several KMT legislators, themselves opposed to the project, signed a petition demanding that the government suspend its construction. Even though the then premier Yu Kuo-hua (俞國華) set aside the issue temporarily by declaring that "the fourth nuclear power plant was in no hurry to be built" (Ho, 2003: 692), the project was never truly halted.

In 1990 the construction of NPP-4 was made official and promoted as one of the major upcoming projects in the "Six-Year National Development Plan." During Hau Pei-tsun's premiership, in 1992, the budget allocated for NPP-4, which had been suspended since 1985 due to some internal strive inside the KMT government, was re-opened and NPP-4 became the symbol of the attempt to regain firm control over society, after the unruly years the country had experienced since the lifting of martial law; at the same time NPP-4 turned into the emblem of the government's attention towards economic development. In 1991 the Ministry of Economic Affairs and its executive body, the Executive Yuan, approved the plan and had the Environmental Impact Assessment passed.

Even though, as mentioned in Chapter 1, back then the instrument of the EIA was still in its infancy, constituting a mere black-box process in which no degree of public participation was tolerated, opposition movements nevertheless did make several efforts to take part in the official EIA proceedings regarding the power plant. In fact, the Taipei County government, which was pro-environment, argued with the AEC and obtained five seats on the EIA committee, even though later on the AEC deprived these five anti-nuclear reviewers of the right to participate in the final decision (Ho, 2004: 244). In 1991, after the government announced that an EIA on the plant had in fact been approved, the Yenliao Anti-Nuclear Self-Defense Association (鹽寮反核自救會 *Yenliao fanhe zijiuhui*),[9] formed in 1988, launched a protracted protest and set up a barricade around the construction site. At the height of Hau's premiership with its radical attitude towards social movements, the fierce battle between protesters and Gongliao locals on one side and governmental and police forces on the other had begun. The previously mentioned (in)famous incident portrayed in Tsui's documentary took place in 1991 as a young military reserve on the side of protesters, Lin Shun-yuan (林順源), accidentally drove over a policemen when backing up with his truck. He was sentenced to life imprisonment, as if the killing were intentional (Tsui, 2004; Arrigo and Puleston, 2006: 177). The CEO of the Yenliao Association, Gao Ching-nan (高清南) was sentenced to ten years in prison. Seventeen others were indicted and found guilty (Ho, 2003: 697).

After the incident, further problems arose concerning the allocation of the budget, which was quickly rising due to the continuous delays in the plant's protracted construction. The KMT, still in control of more than half of the seats in the Legislative Yuan, succeeded without problem in securing approval for the yearly budgets for the construction of NPP-4 from 1995 to 1997. The opposition declared that it would continue fighting against the allocated yearly sums in the following years. In 1999, after the Atomic Energy Council approved the license for the construction of the plant, the government revoked the fishery rights of Gongliao locals without any previous notice. At that point residents went to the Executive Yuan to protest against this decision (Tsui, 2004).

Ten days before the presidential elections of 2000, future President Chen Shui-bian (陳水扁) appeared in Gongliao and promised local residents that, if elected, he was going to stop NPP-4. After he took office however, he did not proceed to immediately halt construction even though it could have been easier to do so back then, as the project was still at an early stage of its construction, at the foundation of the first reactor (Ho, 2003: 684). Due to mounting protests on part of anti-nuclear activists, he appointed a commission of 18 experts and industry representatives charged with the task of making some recommendations on how to proceed (Arrigo and Puleston, 2006: 178). Anti-nuclear activists were granted seven seats but no financial allowance for research, nor any authority to pressure Taipower into sharing relevant documents. Afterwards, barely five months after winning the presidency, the DPP made a reckless move to terminate NPP-4, without prior consultation or approval from the Legislative Yuan (Ho, 2005b: 412). The scrapping of the power plant was bitterly attacked in corporate circles and by sections of the KMT, in large part due to Taiwan's mounting economic troubles; being nearly 30 per cent completed, NT$40 million had already been spent on the project (Jacobs, B., 2012: 176).

Trying to "mend fences" with the KMT, Chen Shui-bian met with Lien Chan (連戰) on October 27 of that same year, to discuss the fate of NPP-4. Lien Chan proposed to continue with the construction while at the same time drafting a plan for a quick termination of the operational activities of the first three nuclear power plants (Jacobs, B., 2012: 177). Two hours after the meeting, the DPP Premier Chang Chun-hsiung (張俊雄) announced that the construction of NPP-4 would be halted, due to the difficulty in solving several thorny issues such as the nuclear waste problem and also stated that Taiwan enjoyed a surplus of electricity and was in no immediate need of another plant. The KMT then brought impeachment charges towards Chen by claiming that, by overturning the previously passed legislation without the permission of the parliament, the president had exceeded presidential authority and violated the constitution. However, this motion was never lodged and on November 25, 2000 the KMT publicly announced that its plans to have Chen Shui-bian impeached had been shelved. This course of action seems to be the result of popular surveys, indicating a slump in support towards Chen but not supporting his impeachment. Chen himself, after apologizing to the KMT, tried to avoid any further escalation of tensions, and appealed to the Council of Grand Justices, Taiwan's highest court, to rule on the constitutionality of the decision to cancel the power plant. The latter released its constitutional interpretation, arguing that the order to halt the construction had not been issued properly, on January 15, 2001 (Ho, 2005b: 412; Lyons, 2009: 65); one month later, construction of NPP-4 resumed (Ho, 2014a: 8).

Rumor has it that DPP lawyers were prepared to battle the verdict of the highest court but that Chen Shui-bian took the decision to continue with the project, right after a meeting with Lee Teng-hui (李登輝), in which Lee's political support for the DPP was exchanged for a continuation of the project (Linda Arrigo, interview, November 3, 2011). On January 31, 2001 the project was finally put up for vote in the Legislative assembly. The motion to continue with the construction was passed with 134 members in favor and 70 members opposed. As noted by Shih Fan-long, "To complicate the matter further, the issue of nuclear power goes beyond domestic politics: considerable economic and political interests of superpowers and multinational companies were involved, affecting the delicate balance of international relations" (Shih, 2012: 303).

The ruling party's change of attitude on nuclear energy was a bitter blow for environmentalists as they could see how the majority of DPP legislators fell silent on the issue of NPP-4 and Taiwan's energetic future almost overnight, despite the "nuclear-free country" slogan being enshrined in their party charter since the year 1986 (Ho, 2003: 694). After more than a decade of sustained struggle, locals were extremely disappointed by the DPP's turnaround on the construction of NPP-4: several community members became demoralized and the whole movement lost stamina (Shih, 2012: 304). Retracing a general pattern of social movements in Taiwan (Ho, 2010a) the local anti-nuclear platform and the Yenliao Association in Gongliao have both enjoyed a new phase of activism since the year 2010. Right after the Fukushima incident, as anti-nuclear activists organized a first protest parade on March 20, 2011, the DPP tried to re-launch its visibility by organizing banners and supporters for the protest. Activists were so angry at what they perceived to be a clear attempt at gaining votes that they coined the slogan "NPP-4 is a joint venture of both the green and blue camp" (核四是藍綠共業 *hesi shi lan lü gongye*) (Tsui Shu-hsin, personal email communication, July 6, 2012) to signal their distance towards both parties, deemed equally guilty for the protraction of this controversial project.

Even though NPP-4's construction is de facto almost completed, anti-nuclear activists' battle throughout the years has nonetheless produced numerous positive results. First of all it has forced the state to rescale its initial nuclear-expansion agenda, even though indignation and anger remain, for what concerns the amount of money spent on the facility. The cost of the project, the biggest of its own in Taiwan, was initially estimated to be around US$6.5 billion (Power Technology, 2012) but it has now grown to US$10 billion. In fact, according to opposition Legislator Tien Chiu-chin, Taipower calculated in 2000 that it would cost NT$80 billion (US$2.6 billion). The estimated cost nowadays is at least four times higher (de Changy, 2012).

A long list of structural and operational problems

Throughout the years a long list of scandals and incidents have marred NPP-4's construction and delayed its operations.

First and foremost anti-nuclear activists worry about the elevated risks associated with an earthquake in a highly seismic country such as Taiwan; in fact, NPP-4 is located in the immediate vicinity of over 50 schools, at a distance of 40 km from the capital, Taipei and within a radius of circa 100 underwater volcanoes, a few still active (Chao, 2012).

Another major problem pertains to the management and operational activities of the plant (Tsai, J., 2011). In fact, while the managing contractor is General Electric (GE), the turbines in the reactors have been contracted to Mitsubishi and other parts of the reactors yet have been subcontracted to Hitachi and Toshiba, which are GE's Japanese partners (Power Technology, 2012). A common fear among anti-nuclear activists is that, given the numerous companies having a stake in the project, should something happen, no one would take responsibility (Hsü Kuang-jung, interview, September 15, 2011; Lai Fen-lan, interview, November 24, 2011).

Furthermore, while the plant was, in its initial planning phase 25 years ago, supposed to be completely computerized, the project has taken so long to build that the sophisticated technology it was based on is by now obsolete; Taipower's changing of the original design has also raised concerns among AEC's experts, with a few officials resigning in sign of protest after Fukushima and the heightened attention towards nuclear safety (Ho, 2014a: 6). Due to the numerous developers involved in the project and the advancements of technology in the past 20 years, several components of the reactors have become difficult to find or simply do not fit with newer parts anymore, causing concern among the public and green activists (Pan Han-sheng, interview, September 1, 2011). Moreover, this is the first plant that Taipower has had to build in its entirety; the company previously restricted its role to the actual management of nuclear facilities after other companies (e.g., General Electric or Japanese companies) carried out the actual construction. Numerous media reports have exposed scandals such as electric cables chewed by rats (Tsai, J., 2011) and my sources have spoken of substitution of critical components of reactors with plastic parts unable to endure exposure to high temperatures (Hsü Kuang-jung, interview, September 15, 2011).

Due to the numerous design and engineering flaws of the fourth nuclear power plant, revealed over the years, representatives from environmental NGOs, local communities and engineering experts established an official commission, named the "Oversight Commission on the Safety of the Fourth Nuclear Power Plant," that passed a resolution in August 2011, which stated that the construction process should be stopped, unless Taipower reforms its engineering procedure before the end of the year 2011. The government ignored this resolution but promised that the safety of the fourth nuclear power plant would be ensured through a peer-review process carried out by international experts. According to two of my interviewees, one a member of the GP and the other of the GCAA, this amounts to a mere façade display on the part of the government (Wang Chung-ming, interview, September 17, 2011; Chao Chia-Wei, interview, October 24, 2011); in fact, by "international experts" the government refers to those from the World Association of Nuclear Operators (WANO) and Nuclear Regulatory Commission (NRC) from the United States, whose credibility should be put in question, they maintain, considering that the Fukushima Dai-ichi Plant had passed its WANO examination with full-fledged colors in 2009, barely two years before the disaster.

Another controversial issue for those who oppose the project is the emergency plan that the government has set up, should a nuclear accident really take place on Taiwanese soil (Liu, 2011: 141–143). First, the evacuation zone has been set at 20–30 km, instead of the 80 km radius evacuation zone recommended by the United States. Even with such a limited evacuation perimeter, a minimum of five million people would have to be relocated in case of an accident (Lee, 2014b: 3). Paul Jobin, himself a scholar of environmental pollution and related health damages, believes that this might indicate a somewhat naïve attitude on part of the authorities, who refuse to face the eventuality of having to resettle more than six million people, in case a nuclear disaster really were to take place at one of the three nuclear facilities located in northern Taiwan (Jobin, 2012: 25).[10]

Taiwanese erotic novelist, turned anti-nuclear activist, Liu Li-er (劉黎兒), writes in her book that no additional safety measure, such as providing local authorities with iodine tablets, equipping them with enough buses for evacuation and being capable of formulating accurate forecasts on wind direction, have been fulfilled under what the government calls its *District Plan for Emergency Management* (緊急應變計畫區 *jinji yingbian jihuaqu*) (Liu, 2011: 142).

A long-time connoisseur of Japan, where she lived for many years until the Fukushima disaster, Liu thinks that just like in the Japanese case, where governmental response and management of the post-disaster situation was characterized by a high degree of inadequacy, Taiwan too is in urgent need of strengthening the country's emergency responses and has to lay priority on fostering a transparent and effective communication with the population. Effective communication and openness can help in shaping the public's perception of a disaster as well as avoid misinformation and distrust; without these elements, any remedy policy would be ineffective (Liu, 2011: 142, 150).

Last but not least trust in Taipower's (namely in the government) capacity of handling a nuclear facility is at an all-time low. A survey carried out by the Atomic Energy Council discovered that more than half of the people residing in the vicinity of NPP-1/2/3 displayed similar levels of trust towards nuclear plants' safety (between 58 percent and 66 percent) with a much lower percentage if the question shifted on the level of trust in the capability of the government to effectively handle a nuclear emergency (between 37 percent and 46 percent) (Atomic Energy Council, 2010). A survey by Ho Jung-chun *et al.* (2013) discovered that 94 percent of Gongliao residents are opposed to NPP-4 and displayed a high level of distrust (also 94 percent) towards the safety management capability of Taipower and the government (Ho *et al.*, 2013: 786). Three different opinion polls conducted in 2013 by both the DPP as well as by pro-KMT polling agencies showed that between 58 percent and 69 percent of the citizens are opposed to NPP-4. The share of those who support the project is between 18 percent and 25 percent (*Liberty Times*, 2013 as quoted in Ho, 2014a: 2).

Further doubts in regards to the need for a fourth power plant in the country arise out of the consistent discrepancies between, on the one hand, energetic shortages emphasized by the government and the pro-development faction (Taipower's White Paper, 2013 as quoted in Wang and Lee, 2013a: 3) and what, according to environmental activists, is in fact a surplus of energy reserves, which Taiwan has accumulated in the past two decades, without the contribution of a fourth nuclear plant (Tsui Shu-hsin, interview, March 24, 2011; Hsü Kuang-jung, interview, September 15, 2011 among others), an issue that will be dealt with later on in the course of this chapter.

At any rate the numerous structural problems of this facility contribute to heighten nuclear safety awareness among the general population, in the aftermath of the Fukushima disaster. The next section will focus on the political fallouts of the Fukushima incident on Taiwan's domestic situation, trying to assess how different stakeholders, from anti-nuclear activists to political parties, have sought to transform their strategies to utilize the nuclear meltdown to their own advantage.

The Fukushima effect and its reverberations

As mentioned above, in the 1990s, due to several factors such as an increase in the maneuver space for activists after the lifting of martial law (1987); the concomitant mass scale reverberations of the Chernobyl incident (1986); and the creation of a green party inside Taiwan (1996), public opinion was quite informed about nuclear issues (Jobin, 2012: 2). In comparison, the following decade (2000–2008) was one of cooptation for green movements, which became "institutionalized" following the Democratic Progressive Party's (DPP) electoral success (Ho, 2005b: 405; Jobin, 2010: 49). With green movements becoming quieter on nuclear energy, citizens were also less informed and thus their perception of risk, considering the lack of serious nuclear accidents at that time, decreased. The Fukushima Dai-ichi event, in all of its drama, has had the effect of giving new stamina to the anti-nuclear power movement globally, while bringing nuclear energy concerns to the fore of all of Taiwan's major parties' political agendas and platforms. In fact, the nuclear meltdown and Japanese people's losses and suffering after the massive earthquake and tsunami that wreaked havoc on the country, have reinvigorated activists' fight against NPP-4. In light of Taiwan's own situation: a seismic, overly populated country with nuclear power plants located side by side with densely inhabited centers and in close proximity to numerous underwater volcanoes (Chao, 2012), it is easy to understand why fear of a nuclear disaster inside the country has been growing. While in the aftermath of Fukushima some people had high hopes for a future nuclear-free Taiwan, after the 2012 presidential elections held in January, newly re-elected President Ma Ying-jeou initially stated that the construction of controversial NPP-4 would go on as planned (Tsai, 2012); at the same time he declared that the shelf-life of the three already existing facilities would not be prolonged. As we shall see towards the end of this chapter this scenario has, meanwhile, radically changed, in the aftermath of another unexpected event and its reverberations for the anti-nuclear cause, the Sunflower Movement.

Having failed to stop its construction, Taiwanese anti-nuclear activists' main goal has shifted to preventing NPP-4 from ever becoming operational. To reach their purpose they employ the powerful imagery of death evoked by the incident in Fukushima with its unforeseeable and long-standing health consequences, spearheaded as an example of what could take place on Taiwanese soil if the plant is ever to become operational (Hsü Kuang-jung, interview, September 15, 2011; Tsui Shu-hsin, interview, September 20, 2011; Lai Fen-lan, interview, October 30, 2011). This strategy is extremely effective; in fact, it has been proved that environmental claims are more successful if protesters "piggyback on dramatic real-world events," by bringing attention to issues that would otherwise be neglected (Ungar, 1992). In the aftermath of a news-magnet event such as the Fukushima meltdown and the increased media attention towards this topic (*Apple Daily*, 2011; Our Island, 2011a, 2011c, 2012c; Tsai J., 2011; Chang, P.W., 2012; Fang, 2012: 8; Lee, 2012a, 2012b, 2012c), nuclear awareness in Taiwan has been growing steadily. Right after the Japanese disaster the leading scientific journal *Nature* rated two of Taiwan's nuclear plants among the world's most dangerous (Butler, 2011: 400).

Given the significance of the nuclear incident and the crucial political moment when it took place – barely ten months before Taiwan's presidential election of 2012 – several concerned stakeholders, from political parties to social activists, sought various methods to exploit it to their own advantage.

As far as Taiwanese activists are concerned, the anti-nuclear movement has experienced a phase of intense activity with countless protests tallying on the fears created by Fukushima. Right after the nuclear disaster, nuclear-related issues got so much attention that they planned a host of different activities to fully exploit the momentum. In March 2011, around 2,000 anti-nuclear protesters demonstrated all over Taiwan for an immediate end to the construction of the island's fourth nuclear power plant. Protesters were also opposed to any lifespan extension for the three other already functioning nuclear plants (*Asia One News*, 2011). Massive protest parades were organized on April 30, 2011 with 5,000 people joining an anti-nuclear protest in Taipei City, holding yellow banners and waving sunflowers as part of a nationwide "No Nuke Action" campaign against the construction of NPP-4 and in favor of more renewable-energy-oriented policies (Lee, 2011a); the turnout was so high that a new phrase was coined to describe these parades: "the 430 movement" (430 行動, 430 *xindong*).

In 2012, rather than declining, the number of protests increased; in March, about 2,000 people staged an anti-nuclear rally in Taipei to commemorate the one-year anniversary of the massive earthquake and tsunami that hit Japan the previous year, while renewing their hopes for a nuclear-free island.

One year later, on March 9, 2013 200,000 people took to the streets in Taipei and three other major cities, again requesting that the construction on the fourth nuclear power plant be halted. The protest in Taipei was the largest ever anti-nuclear demonstration held in Taiwan to this day (Taiwan Insights, 2013).

Afterwards the cultural community organized a series of anti-nuclear events in Taipei, under the name "Movement to say no to nuclear plants number four, five and six" (不要核四、五六運動). This anti-nuclear campaign, which began on March 15, 2013, included in its ranks several cultural heavyweights, including film director Ko I-chen (柯一正) and writer Hsiao Yeh (小野) (Tseng *et al.*, 2013) and was aimed at bringing individuals opposed to nuclear energy every

Figure 3.4 Sunflowers as a symbol of clean energy were already being used at an anti-nuclear rock concert, November 5, 2011, Taipei (source: Simona Grano).

Friday to Liberty Square, where activists and intellectuals held anti-nuclear speeches and sought to raise awareness of the underlying risks connected to nuclear power plants and more specifically to the faults and dangers of NPP-4.[11] Three years after the Fukushima incident protesters are still active and make use of every auspicious event, trying to spin it in their favor. A strong earthquake that hit northern Taiwan on February 12, 2014 has renewed concerns about the possibility of one of the nation's plants being damaged by a quake. This earthquake was particularly worrisome due to its magnitude (4 on Yangmingshan) and the fact that the epicenter was in Taipei's Shilin District (士林區) (Lee, 2014a: 1). On February 16, 2014 protesters took to the streets in what was termed "an anti-nuclear run"; this event was meant to symbolically portray the run Taiwanese would have to make if something real were to happen to one of its plants (Lee, 2014b: 3). Barely six days later, an eight-hour long concert, supporting the abolition of nuclear power, was held in Taipei (Lee, 2014c: 3). Such events are principally meant to encourage young people to join anti-nuclear campaigns and become more involved in social movements' activities.[12]

If activists have sought to employ the Fukushima disaster to their own avail, using protests and parades as their main tools, major political parties have done the same, by incorporating nuclear energy issues and concerns into their electoral platforms, to avoid losing out to the competition.

While several international governments, from Germany declaring that the country would end its reliance on nuclear energy by 2022, to Switzerland and Italy, have shelved their plans for nuclear expansion (or initiation in the Italian case) shortly after the Fukushima disaster, pre-electoral campaigning in Taiwan all throughout 2011 reflected a division according to political orientations and so did a few opinion polls carried out after Fukushima among the general population;[13] the *Liberty Times* (自由時報 *ziyou shibao*) suggested predominantly anti-nuclear feelings among its readership, with more than 87 percent asking for a re-examination of NPP-4, as requested by the DPP;[14] the *China Times*, sympathetic to the KMT, reflected opinions on the other side of the political spectrum with a total of 57 percent of the people agreeing with President Ma's stance on nuclear energy (as quoted in Jobin, 2012: 12). A more neutral position is attributed to the *Apple Daily* with 56 percent of its respondents wishing to stop NPP-4's construction, 32 percent wishing to continue with the project and 12 percent with no opinion (*Apple Daily*, 2011).[15]

As noted by Paul Jobin, while this figure is not an absolute majority it does nevertheless show that more people have become anxious in regards to nuclear power generation after Fukushima.

As for politicians, the KMT's official stance towards nuclear energy since 2011 has been evolving according to the changes that have taken place in public opinion. Initially, right after Fukushima and during the electoral campaign for the elections of 2012, Ma Ying-jeou spoke about de-commissioning the first three nuclear power plants by 2025, while continuing with the construction of controversial NPP-4; this meant that Taiwan was going to become nuclear-free at the earliest in 2055; a position that is reminiscent of the previously mentioned proposal by Lien Chan in 2000 (Jacobs, B., 2012: 177). President Ma was keen to reassure Taiwanese that: the new plant would begin commercial operation only under secure conditions; that NPP-1 would soon be decommissioned; and that the lifespan of NPP-2/3 would not be prolonged.[16] He insisted that Taiwan can only take a gradual approach towards the achievement of a nuclear-free homeland because of energy security concerns, the need to keep electricity prices reasonable and the goal to reduce – or keep in check – greenhouse gas emissions. On this point, Stephen Shu-hung Shen, previous minister of the Environmental Protection Administration (2008–2014), is famous for his opinion that the immediate abolition of nuclear energy is a "romantic idea" but highly impractical as well as in conflict with Taipei's goal of meeting its carbon reduction targets and reducing air pollution (Chu Yu-chi, Director General at EPA, interview, November 2, 2011). In fact, Minister Shen estimates that if plant n. 4 were to be canceled and replaced by coal-fired plants, Taiwan's carbon emissions would soar. At any rate, considering that in 1985 power generation derived from nuclear accounted for 52.4 percent (Taiwan Statistical Data Book, 1996: 94), in 1990 Taiwan nuclear energy accounted for 36 percent of its installed capacity of electricity output (Ho, 2014a: 4) and the share dropped to 19 percent 20 years later, one can observe how the country has moved along the trajectory of trying to diminish its dependence on nuclear energy.[17]

A few months after winning the presidential race in 2012, Ma declared that nobody inside Taiwan is opposed to the government's policies towards nuclear energy (沒有人]反對台灣政府的核能政策 *mei you ren fandui Taiwan zhengfu de heneng zhengce*) (PNN, 2012). In response, anti-nuclear groups collected 15,000 signatures, composed a huge banner with the names of the signatories and brought it to the Presidential Palace to confront Ma Ying-jeou, chanting the slogan: "I am someone, I oppose nuclear energy" (我是人, 我反核 *wo shi ren, wo fan he*)[18] (Ma, 2012).

Concerning the DPP's position towards nuclear energy and its electoral platform, after the Fukushima disaster the party reverted to its early anti-nuclear stance with several politicians competing to emphasize their opposition to nuclear power and NPP-4, in order to gain visibility in the year preceding the upcoming presidential elections of 2012.

As mentioned above, Dr. Tsai Ying-wen, the then presidential candidate for the DPP, had tried to turn the issue of ending Taiwan's reliance on nuclear energy into the *pièce de résistance* of her campaign, with the much-heralded slogan of achieving a "nuclear-free homeland" (非核家園) by the year 2025. Ironically, her candidacy was announced exactly on March 11, the very day of the earthquake that hit Japan. After the nuclear meltdown she made several important and audacious statements on the nuclear issue.

However, while she officially supported several protest rallies held against nuclear energy, such as the already named "430 movement" held in Taipei on April 30, 2011, she preferred not to join these parades. Such decisions, coupled with her somewhat ambiguous attitude towards NPP-4, supporting its construction but not its operation (Shih, 2012: 297; Ho, 2014a: 14), confused and angered the most radical wing of her electorate as well as those hard-core, no-compromise green activists, staunchly opposed to the plant, who dubbed her behavior as "opportunistic" and "equivocal" (Tsui Shu-hsin, interview, March 24, 2011; Linda Arrigo, informal communication, November 3, 2011). Many environmental activists perceived her position as a calculated move to take advantage of the high tide of opposition towards nuclear energy, generated by the nuclear disaster in Japan. The DPP's prior betrayal of its pro-environmental commitments was still fresh in the minds of anti-nuclear activists. At any rate, the schism between the "older" and the "newer" generation of anti-nuclear activists has become even more deep-seated after 2011, with TEPU and other old-guard groups still leaning towards a cooperation with DPP and newer groups, such as GCAA, taking a clear stand in favor of more independence and distance from both political parties, equally guilty of having betrayed environmental protection. Therefore, many green activists chose to cast their vote, whenever possible, with the Green Party, which ultimately did not manage to exceed the 5 percent threshold required to get a seat in the Legislative Yuan for any of its ten candidates.[19]

In her defense, Tsai maintained that she did not agree to press for an immediate halting of the construction mainly in consideration of the legal conflict that would have arisen with the developers (Jobin, 2012: 9). Tsai's explanation was that the costs of compensation for a breach of the construction would have turned out to be higher than finishing the project and maybe coming up with an alternative solution for its use.[20]

As stated by Paul Jobin, "Once again, Taiwanese nuclear politics were back to the situation of the 2000s, presenting a strong antagonism between blue and green camps, with NPP-4 as a cornerstone" (Jobin, 2012).

In fact, the nuclear energy controversy proves once more the depth to which environmental issues become politicized in Taiwan, with citizens dividing along the pan-green or pan-blue positions.

On this point it is important to note that while most of Taiwan's citizens are more aware and concerned about a potential nuclear accident hitting the country after Fukushima, when asked to pay higher prices for energy, in favor of a "nuclear free home," few declare themselves willing to forsake nuclear. In the pre-Fukushima era, an island-wide survey carried out by Taiwan Environmental Protection Union (TEPU) in 2000 discovered that 42 percent of the people thought nuclear should be gradually phased out; 32 percent thought of it as "safe and necessary" for the country's energetic survival while 21 percent considered it "unsafe but necessary" (as quoted in Williams and Chang, 2008: 82–83), thus clearly illustrating the ambivalent feelings displayed by many Taiwanese in regards to nuclear energy, even more so in the pre-Fukushima period. In fact, we have to keep in mind that the debate on nuclear energy touches upon practical concerns, such as the costs to be sustained if the country really were to move away from its nuclear trajectory and the ensuing problems that would arise, to ensure a steady supply of resources, lest risking a rise in electricity prices, which might create resentment in the population. For activists this attitude is a clear sign that Taiwanese citizens are quite receptive to the much-propagandized problem of the island's scarcity of resources. In fact, the notion that the country's energetic output is not adequate for current needs is an issue that concerns many. In Pan Han-sheng's words: "Under the state pro-growth ideology, the campaign to promote nuclear energy as a leading state sector has been successful in marginalizing opposite view, as emotional and lacking any scientific evidence" (Pan Han-sheng, interview, July 1, 2011). To those who emphasize the country's need for energy, green activists respond that Taipower spreads misleading information by highlighting Taiwan's lack of sufficient energy resources when the country's energetic supplies are actually in excess (Tsui Shu-hsin, interview, September 8, 2011; Lai Fen-lan, interview, November 24, 2011). They emphasize that their proposals are viable without any radical action, such as forsaking air-conditioners and electric illumination.

In Tsui Shu-hsin's words:

> We only call for reasonable use. We need smarter policies, such as raising electricity costs and energy taxes. These measures would be directed at the industrial sector and not at private households. Even if all three operational nuclear power plants were to cease operation, our back-up capacity would still remain at about 20.9 percent. It is a myth that nuclear power is indispensable. Taiwan's electricity resources are sufficient without nuclear power; the so-called supply crisis is premised on huge economic growths in future decades which will not lead to insufficient supplies but to a reduced back-up capacity, at perhaps above 10 percent. We need to look past these threats.
>
> (Secretary-General of the Green Citizens' Action Alliance and member of the NNP4 Safety Committee, GCAA's headquarters, September 8, 2011)

Green Citizens Action Alliance has come up with a three-pronged alternative plan to cut energy consumption: (1) reducing the ratio of electricity-thirsty industries; (2) increasing the generation of renewable energies; and (3) increasing energetic efficiency. GCAA's members believe that if all of these three provisions were to be implemented, the energy saved would be equal to the production capability the country could derive in the future from NPP-4 (Tsui Shu-hsin, interview, September 20, 2011).

Environmentalists believe that a partial solution could be found in the increase of energy costs, which are currently too low; a rise would encourage people not to waste energy and be more conscious about its use (Pan Han-sheng, interview, September 1, 2011; Tsui Shu-hsin, interview, September 8, 2011; Hsü Kuang-jung, interview, September 15, 2011; Lai Fen-lan, interview, November 24, 2011). In Hsü Kuang-jung's words:

> Taiwan is not short of power, because the amount of electricity generated by nuclear power accounts for only 19 percent of the total power supply, and can be replaced by improving electrical transmission efficiency or energy-conservation measures.
>
> (Hsü Kuang-jung, Professor of Atmospheric Sciences, NTU, September 15, 2011)

In this regard it is has to be mentioned that energy prices in Taiwan have remained steady for more than 20 years, from 1983 to 2006 (Shih, 2012: 310), a fact that has been heralded by pro-nuclear bureaucrats as possible thanks to the country's domestic generation of nuclear power. A recent increase of electricity rates, unilaterally announced by Taipower, provoked fierce reaction among opposition forces. DPP legislators claim that before resorting to such a raise – the third in less than a year and a half – the company should address those issues that have increased the company's expenditures in the past few years, such as prices of favor accorded to commercial clients and the high extra costs sustained for the prolonged construction of NPP-4 (Wang, 2013b).

Anti-nuclear activists maintain that the government's plan, targeted at changing people's habits, is fundamentally flawed and that what really needs to be done is to reform the pricing mechanism so that those who consume vast amounts of electricity (namely factories) will pay more per unit than those who consume small amounts (households).

> At present, the pricing system favours consumers with big volumes. In reality the public de facto subsidizes the industrial sector's power usage at the cost of NT 2,000 per household. Not only is this policy unjust, it also leads to oversupply and wastefulness.
>
> (Tsui Shu-hsin, September 20, 2011)

Critics of nuclear energy also question the need for a fourth nuclear power plant, considering the small percentage of electricity supply derived from nuclear, which accounts for a mere 18–19 percent (Mathews and Hu, 2013). Taiwan has long been toying with the idea of increasing its share of renewable energies such as solar photovoltaic (PV) and wind turbines.[21]

The storing of radioactive wastes

No matter what the future holds in store for Taiwan's nuclear energy policy, a big headache for governments dealing with nuclear production is what to do with its unwanted by-products, namely nuclear wastes. This issue takes on a whole new dimension in Taiwan where, throughout the past two decades, it has become intermingled with international, racial and environmental-justice concerns. In fact, following Taiwan's complex political situation, as a sovereign state with no international status, deprived of a seat in the United Nations, the Republic of China cannot use the option of reprocessing its spent fuel abroad (Liu, 2011: 145). Therefore, since its first nuclear reactor started being operational in 1971, wastes have been stored either inside its nuclear facilities or, after 1982, on the already mentioned Orchid Island (蘭嶼 *Lanyu*), inhabited by the aboriginal Yami (or Tau in their own language) Tribe.[22]

In the 1970s, when the KMT regime decided to locate a nuclear waste facility on this offshore island it deliberately omitted to provide locals with the truth, leading residents to believe that a fish-canning factory was going to be built on their island (Huang *et al.*, 2013: 1559). The factory was propagandized as an invaluable source of new job opportunities for a low-income community such as

Lanyu. Shipping of the wastes began in 1982; from 1982 to 1996, 100,000 barrels of nuclear wastes were brought to the island (Enn, 2012: 152). While the Yami originally ignored what the government had built (Huang *et al.*, 2013: 1561), after an environmental NGO found out the truth, they began refusing further shipments of wastes and have consistently negotiated for the removal of the storage which, according to the government, was supposed to be a temporary deposit in the first place. Unverified rumors of stillborn infants and birth defects on Orchid Island have increased exponentially although very few medical studies support these statements in a consistent manner until today (Arrigo *et al.*, 2000; Peter Wushou Chang, informal communication, August 21, 2011; Boris Voyer, telephone interview, November 16, 2011; Loa, 2012c). Poor management on the part of governmental bureaus and Taipower in regards to the nuclear-wastes facility on Orchid Island, where numerous leaks have caused radioactive material to seep into nearby sea and land, increases public mistrust towards authorities regarding nuclear technology-related issues (Chung, 2005; Loa, 2012b, 2012c, 2012d; Grano, 2014a).[23]

Even though opinions among residents of Lanyu are not uniform, with some valuing the improved infrastructure and lifestyle opportunities that the new facility has created and some utterly opposed to it, in the past few years and especially since 2011, locals have upped their requests to have the repository moved elsewhere (Loa, 2012d). The Fukushima Dai-ichi incident has given them new strength to renew their protest to have the storage site moved in February 2012; these rallies have been given a high degree of visibility by the local media, especially by those newspapers that are not aligned with the KMT, such as the *Taipei Times* or the *Liberty Times* (e.g., Loa, 2012a, 2012c, 2012d).

In order to avoid the process of "wastes dumping" being associated with "environmental racism" (due to the fact that Orchid Island is predominantly inhabited by the already mentioned Yami tribe of Austronesian descent), Taiwan has been toying (albeit with no success) with the option of "exporting" its wastes abroad. A number of options, all with severe repercussions on Taiwan's international status, have been under consideration. Negotiations went on for a while with China, North Korea and Russia. In 1997 and 1998 Taiwan actually signed agreements with North Korea (Schafferer, 2001: 109); however, after South Korea voiced concerns in regards to North Korea's potential handling of uranium-enriched material, Taiwan's international reputation suffered a great blow and accusation of "environmental classism" (a rich country which exploits a poorer one for its own sake) were repeatedly raised so that Taiwan was forced to abandon the North Korean option (Hsiao, 1999: 51).

Such issues highlight once more the international dimension of the nuclear controversy; in fact, all of Taiwan's six reactors are imported from US vendors – one of the speculations behind the prolonged continuation of NPP-4 being US pressure – and, as shown, Taiwan itself has consistently tried to find a suitable *escamotage* that would allow it to export its undesirable products, "dumping" them onto less developed countries.

Confronted with the Yami's refusal to accept any more barrels and with the nuclear issue having gained visibility among the general population, the government was forced to find alternative storing sites. Finding a suitable repository has been a consistent headache for the Taiwanese government since at least 1999, when it promised to remove all wastes from Orchid Island by the year 2002 and store them in a purpose-built storage facility on the mainland, to be located either in Pingtung (屏東) or Penghu County (澎湖縣). In fact, in 1999 the Ministry of Economic Affairs (MOEA) announced that two sites for storage had been found, one being Da-Ren Town (達仁鄉) in Taitung County (台東縣) and one being Wang-An Town (望安鄉) in Penghu County (Huang *et al.*, 2013: 1562). However, since the AEC drafted the "Act on Sites for Establishment of Low Level Radioactive Waste Final Disposal Facility" in 2006 (AEC, 2006), which sketches the basic framework both for what concerns the criteria to site such facilities but also for incorporating public participation in the decision-making process, the whole procedure has become more complicated, subject to the several interests of the many concerned stakeholders. The act lists a number of conditions that have to be fulfilled for such a facility to obtain a green light (e.g., avoidance of areas where there are active geological faults) and furthermore establishes that an ad hoc committee should be charged with the responsibility of acting as a "watch-dog" over Taipower's siting decisions (Huang *et al.*, 2013: 1562). This committee shall be composed by 17–21 representatives of relevant governmental agencies, experts and scholars (experts having to be at least three-fifths of the total members). Furthermore, the act states that after a suitable site is found, a referendum must be carried out and only in case it successfully passes, can the site be listed as a candidate for hosting the storage complex. Financial compensation matters are also regulated in detail in the act.

Due to the increased knowledge on the part of protesters in regards to nuclear energy and its by-products, locals in Taitung and Penghu have vehemently protested the government's proposal and have set up barricades and raised compensation issues, which had to be dealt with before any decision could be made. Therefore the initial plan, which was to complete the facility by 2013 (Williams and Chang, 2008: 2013), as of 2014 has yet to be defined in both its location and scope and even though the government maintains that it has identified a suitable parcel of land to accommodate the surplus of nuclear wastes, the truth is that, up until now, not a single municipality has agreed to accept such a facility.

Finally, an important but often forgotten dimension of the nuclear energy controversy is related to secondary issues, such as radiation contamination of food products, which takes on a whole new dimension in Taiwan, a country where food safety issues have consistently been on the front page of media reports for the past couple of years, and especially since the 2011 plastifiers scandal. While in no immediate danger of being exposed to radiation plume, Taiwan deals with the fear of contaminated food and radioactive products that may be imported from Japan, the products of which had always been considered as "up-scale" delicacies, devoid of the numerous scandals that so often marr Chinese and Taiwanese domestic goods, and thus a synonym for "safety" and "quality."

Strong reactions among the population and concerned activists were ignited after the announcement of the Ministry of Health in July 2012, which called for a lowering of the safety standards for the contamination of radioactive elements (Our Island, 2012d). The proposal called for increasing the tolerance level for food products, extending it to 300 or even 370 Bq/kg[24] when the international standard states that anything over 100 Bq/kg is considered as "nuclear waste" (Paul Jobin, email communication, October 24, 2012). As of November 2014, only vegetable, fruit, aquatic products, baby formula, dairy products and water imported from Japan require a radiation detection report, with the exception of all food products from Fukushima, Ibaraki, Tochigi, Gunma and Chiba prefectures, which have been banned since the Fukushima Dai-ichi nuclear disaster in March 2011. However, following the latest food-safety scandal of September 2014 the government has been compelled to enforce stricter regulations to avoid any more problems in the future. Thus, the Food and Drug Administration, starting from 2015, will require Japanese companies intending to import teas, candies or cookies to submit a radiation detection report issued by a local authority (Hsu, S., 2014).

The nuclear referendum issue

The first two referenda on the island's fourth nuclear power facility were held in 1994 by the Taipei County government, at the time controlled by a DPP incumbent; they were declared invalid for lack of sufficient turnout. Soon after the Taipei City Government, at that time also under DPP's rule, held its own referendum on the nuclear issue, jointly with the March 1996 Presidential Elections (Hsiao, 1999: 39). The turnout indicated that 52 percent of voters opposed the construction of NPP-4. A fourth referendum was held in 1998 in Ilan County also under a DPP leadership (Ho, 2014a: 7).[25] Finally, in 2001, after the unsuccessful and clumsy attempts to scrap NPP-4 from the country's political future, the DPP government promised the electorate that a national referendum on the issue would soon be held. After the referendum did obtain a legal status in 2003 the DPP proposed that one be held in conjunction with the 2004 presidential election (Ho, 2005b: 413), the focus of which was going to be on China's missile threats. While the DPP has to be credited with giving the instrument of the referendum a legal status, its promise of holding an anti-nuclear plebiscite was never honored (Ho, 2014a: 9). In fact, when Taiwan's first legal referendum was finally held, on the same day of the 2004 presidential election, the issue of nuclear energy was not included among the various issues on which citizens were called to give their opinion.

For almost a decade the nuclear referendum issue lay forgotten until early 2013, when common citizens, Taiwan observers and scholars alike were confronted by the insistent news, circulating on several media outlets, of an imminent government-backed referendum on NPP-4. In reality it was actually the idea of the DPP, which in January 2013 had launched a signature campaign to hold a referendum on NPP-4 that should have taken place in conjunction with the municipal elections of November 2014. The choice of holding the referendum concurrently with a national election was made out of consideration for the high threshold set by the *Referendum Act* (公民投票法 *gongmin toupiaofa*),[26] so as to take advantage of popular mobilization for other political issues (the national election). Since the DPP announced its decision without previously seeking consent from anti-nuclear activists, rumors began circulating that the party was merely trying to increase its electoral success by exploiting the nuclear issue. Therefore the DPP abandoned the referendum plan, which was revived one month later by the KMT. On February 25, 2013 Premier Jiang Yi-huah (江宜樺) suggested using a popular plebiscite to decide on the fourth nuclear plant. The KMT was facing an internal strife in regards to nuclear energy, with several political figures openly expressing their disapproval of NPP-4; among these former Taipei Mayor Hau Lung-bin (郝龍斌) (Ku *et al.*, 2013; Mo, 2013) – by no means a friend of environmental activists – and New Taipei City Mayor Eric Chu (朱立倫), both openly concerned about NPP-4.[27] The government's hope in proposing the referendum was that it could, once and for all, settle the plant's fate (Fell, 2013) thus signaling a departure from the party's traditional, pro-nuclear orientation. The reasoning behind the government-backed referendum is that the public should be called in to settle a thorny issue that politicians haven't been able to solve in more than two decades (with numerous repercussions dealing with safety) and give their opinion on the feasibility of completely forsaking nuclear in the future, while at the same time maintaining enough energy reserves and reasonable costs (Mathews and Hu, 2013).

To Taiwan watchers and scholars, both parties' behaviors appeared as somewhat "out of character." After having advocated for a referendum on the nuclear issue for more than ten years, activists and DPP members, rather than being pleased, disagreed with the way the question about the fate of NPP-4 had been phrased by the government. In their opinion, the question was in favor of the proponent (KMT), being phrased as such: "Do you agree that the construction of the fourth nuclear power plant should be halted and that it not become operational (你是否同意核四廠停止興建不得運轉?)." DPP members openly accused the government of deliberately phrasing the question in a negative way, in order to win the referendum, taking advantage of the structural flaws of the referendum law; this way, citizens who chose not to participate in the referendum would be considered as supporting the construction of the plant. Instead, the DPP advocated for an immediate halting of the plant's project; knowing full well that the KMT would never agree to do so, the party argued that the government should at least provide a report on the quality of the construction to prove that the plant was worth the investment needed to complete it. The truth is that, after having advocated for a referendum the DPP abandoned the idea only to walk into a trap set up by the KMT with its own referendum proposal. In fact, if the DPP refused to be engaged it was going to risk political credibility having previously advocated for a referendum itself. Therefore the party opted for a two-pronged strategy: mobilizing its supporters' base while at the same time attempting to pressure the KMT into revising the *Referendum Act*, by lowering the threshold for referenda to be declared valid (Shih, H.C., 2013). After the KMT's proposal to settle this two-decade-long controversy via a popular plebiscite, those few KMT members who had openly expressed their opposition to the construction of NPP-4, sided with the official party line, claiming that it represented a chance to let popular will decide. Nevertheless, a few discordant opinions remained. According to the *Taipei Times*:

> Lin Tsung-yao (林宗堯), a former member of the Atomic Energy Council's Fourth Nuclear Power Plant Safety Monitoring Committee and a former General Electric engineer, who the KMT had been expecting to come to the rescue of the plant project, said that the plant is a hopeless case as far as safety is concerned.
>
> (Fang, 2013: 8)

This would indicate a certain uneasiness in KMT circles, in regards to NPP-4 becoming operational.

Whether the KMT's proposal stands to indicate that the party is becoming increasingly attentive to people's wishes or whether the Nationalists have simply realized that the alienation of an increasing number of individuals who oppose nuclear energy could backfire and threaten its position as leading party, remains to be seen. A good old fist fight, reminiscent of the early days of Taiwanese democracy, broke out in the parliament on August 2, 2013, pitching anti-nuclear as well as DPP supporters against KMT legislators (Tomlinson, 2013), with requests (or rather yells and screams) for the government to *first* amend the *Referendum Act* and *then* talk about NPP-4 (先修公投, 再談貢寮 *xian xiu gongtou, zai tan gongliao*).

However, the referendum plan was postponed indefinitely in September 2013 by the very same KMT legislator, Lee Ching-hua (李慶華), who had first launched the idea. Afterwards, the nuclear referendum issue lay forgotten, in the midst of more pressing concerns and protests, such as those towards the Free Trade and Service Agreements, until April 2014.

Tallying on the high degree of popular support towards the Sunflower Movement (Wang, 2014), anti-nuclear activists resumed their battle to lower the referendum threshold, criticizing the government for what they considered to be a hypocritical stance in regards to the quorum for a referendum to have legal validity. In fact, while the KMT continues to uphold the 50 percent threshold in regards to NPP-4, it has allowed a casino referendum, held on Mazu in 2012, to be declared valid with a 40 percent turnout of voters, thus with a simple majority of votes rather than being based on shares (Shih H.C., 2013).[28]

As mentioned above, the Sunflower Student Movement took place between March and April 2014 in Taiwan and was mainly focused on requesting a clause-to-clause review of the controversial CSSTA. However, given the fact that several of the students making up the core component of the Sunflower Movement are also anti-nuclear activists, as shown in several anti-nuclear leaflets distributed during the three-week-long protest, it is correct to say that social activists in contemporary Taiwan aspire to a common goal, which transcends old categorizations and divisions and is firmly rooted in the upholding of Taiwan's

Figure 3.5 Anti-nuclear activists at the Sunflower Student Movement protest, April 17, 2014, Taipei (source: Rosa Enn).

democratic principles and the protection of its citizens' independence and right to self-determination, from both the government as well as from Mainland China, no matter the formal cause they stand for. In fact, other than just advocating for an oversight mechanism, which could monitor future cross-strait negotiations, the student leaders of the Sunflower Movement established a new organization, *Taiwan March* on May 18, 2014, which aims to reform Taiwan's referendum law and push for a legislative review of the CSSTA, along with other cross-strait pacts and economic bills. The direct link between the Sunflower Movement and the anti-nuclear power movement is also visible in the public declaration of one of the students' leaders, Chen Wei-ting (陳為廷), who was very active on social media in defending an anti-nuclear rally where activists (including himself) occupied Zhongxiao West Road, blocking the traffic for hours, and were criticized by many annoyed citizens who had to be re-routed due to the parade (Tsai and Chung, 2014: 3).[29]

However, even more important for NPP-4 and the anti-nuclear cause was Lin Yi-hsiung's[30] hunger strike to protest against the construction of NPP-4, which started on April 22, 2014. His actions, which seemed even more dramatic given his failing health, are credited with launching a series of anti-nuclear rallies all over Taiwan (Lee, 2014d), which pressured the government, weakened by the two-week-long Legislative Yuan occupation and by the increasing number of protests of the past few years, into releasing a statement on April 27, 2014; in it the Ma administration announced that it was halting work on the Lungmen nuclear power plant (the already mentioned NPP-4), pending a referendum on its future and that fuel rods would not be inserted before the referendum. Specifically, the KMT agreed to seal off NPP-4's n. 1 reactor after a series of safety tests are completed and halt all work on reactor n. 2 (Hsu, J., 2014). So far, the plant has cost more than US$9.9 billion – NT$274 billion (World Nuclear Association, 2012; Ho, 2014a: 5).

Whether such a popular plebiscite will ever be held is difficult to say. At any rate, in July 2014 TEPU members began collecting signatures for initiating one and have managed to surpass the threshold (in terms of signatories) needed (Lai, 2014b). The fact that the decision was taken only a few months prior to the November 29, 2014 municipal elections, considered as a prelude of the 2016 presidential elections, and thus of extreme importance for all parties, cannot be underestimated.

Conclusions

Taiwan, with its fast-paced race from a pre-industrial society to a post-industrial one, provides interesting examples of how novel kinds of risks, brought about by modernization, become part of everyday life. As shown by relevant data collected through interviews and written sources, the response to the Fukushima disaster took different forms in Taiwan with no homogenous attitude and a division that often traces political orientations.

The numerous scandals and "incidents" surrounding NPP-4's lengthy construction and the several already mentioned rumors pertaining to the nuclear-wastes' facility on Orchid Island contribute to lower local residents' trust towards the capability of Taipower of managing the plant effectively and transparently. Such scandals, to which media outlets politically close to oppositional parties have given high visibility, have contributed to a rapid decline of social trust, with far-reaching social and political ramifications. Nevertheless, the fact that despite all this, the decision to freeze NPP-4 was taken due to pre-electoral concerns (the municipal elections of November 2014), clearly illustrates that "political" interests are the main drivers behind strategic decisions, which deal with energy concerns and a powerful national industry, whose monopoly resides with the state and is linked to the issue of national sovereignty. To once more underline the importance of the year 2011 as a watershed year, we have witnessed how, from an unfavorable situation of "party dependence," in which anti-nuclear activists had no choice but to rely on the DPP's political strength, after the nation-wide impact generated by the Fukushima disaster, they have gradually become "independent" and have freed themselves from party constraints; something made possible only due to the renewed interest of the media and the public towards nuclear issues worldwide. While political parties initially found themselves in a position of "superiority," it is now they, as witnessed for example with the KMT's attempt to exploit the referendum issue to its own advantage, which have to try to incorporate environmental concerns in their electoral platforms so as to gain visibility, meet popular demands and attract voters. Prior to the Fukushima incident, except for a minority of anti-nuclear activists, neither the government nor society at large gave much thought to the issue of nuclear energy.

Therefore, the biggest change brought about by the Fukushima disaster in 2011 and by the re-awakening of anti-nuclear sentiments triggered by the Sunflower Movement protest in 2014, is the crossing of the partisan divide; a change that has rendered the anti-nuclear movement more mainstream, globalized and rejuvenated, leading it to break its ties with the traditional party orientations that had characterized Taiwanese political life (from environmental to other issues) for more than two decades. Although it is quite safe to assume that Taiwan will not radically depart from its nuclear trajectory in the near future,[31] the Fukushima incident first and the most recent protests and dissatisfactions towards the government have had two effects: they have forced many previously indifferent citizens to be confronted with the risks the country would face, in the case of a nuclear catastrophe; and second they ignited strong social protests and concerns, which the KMT government could no longer ignore. Furthermore, after the prolonged and fierce debate surrounding the issue of the fourth nuclear power facility, it is unconceivable that any government, blue or green, will ever dare to propose a fifth facility, de facto putting an end to the KMT's initial nuclear expansion plans, symbolizing the anti-nuclear movement's greatest victory.

Even better suited to understand the contours of environmental politics in Taiwan than the nuclear energy controversy is the story of the petrochemical industry, the damages of which in terms of air, water and soil pollution are geographically spanned out across different regions of the country, as we shall see in the next chapter.

Notes

1 When Chiang Ching-kuo, son of Chiang Kai-shek, served as the premier of Taiwan in the early 1970s, he announced that the government would start a series of infrastructure projects. They are known as the "Ten Major Construction Projects" (十大建設 *shi da jianshe*), which laid the groundwork for Taiwan's economic miracle in the 1980s. The development of the nuclear energy and petrochemical sectors were among them.

2 A first pact between the United States and Taiwan to ensure supplies of nuclear fuels for Taiwan's power plants was signed between the Taipei Economic and Cultural Representative Office in the US and the American Institute in Taiwan in 1972 and amended in 1974. The pact expired on June 22, 2014 but a new agreement to ensure the continuation of such services had already been finalized in December 2013 (*Taipei Times*, 2013c).

3 NPP-1, which carries the name of Chinshan plant, is actually located in the town of Shimen, right at the border with the town of Chinshan (金山); thus some scholars talk about Shimen while others about Chinshan when referring to the NPP-1's location.

4 Taiwan currently has 4,927 MWe of nuclear power capacity by means of three active plants and six reactors (data retrieved from the World Nuclear Association on September 5, 2014).

5 The boiling water reactor (BWR) is a type of light water nuclear reactor used for the generation of electrical power. It is the second most common type of electricity-generating nuclear reactor after the pressurized water reactor (PWR), also a type of light water nuclear reactor. The main difference between a BWR and PWR is that in a BWR, the reactor core heats water, which turns to steam and then drives a steam turbine. In a PWR, the reactor core heats water, which does not boil. This hot water then exchanges heat with a lower pressure water system, which turns to steam and drives the turbine. The two reactors in NPP-4 belong to the third generation of advanced boiling water reactors; the first of their kind in Taiwan (Power Technology, 2012).

6 National Taiwan University (國立臺灣大學 *guoli Taiwan daxue*) is considered as the "liberal" university, *alma mater* of numerous social activists and concerned environmentalists, as opposed to National Chengchi University (國立政治大學 *guoli zhengzhi daxue*), whose reputation as a more "conservative" nationalists' cadres school is well known. Shelley Rigger in *Why Taiwan Matters: Small Island, Global Powerhouse* calls Chengchi University "a training ground for future KMT leaders" (Rigger, 2011: 63).

7 GCAA was initially, back in 1992, the Taipei Branch of TEPU (Ho, 2014a: 10; Tsui, Shu-hsin, interview, September 8, 2011). The increasingly fierce antagonism with the national federation led to a secession in the year 2000 and GCAA was finally established. The two organizations are different in their composition; while TEPU's core is still made up of scholars, intellectuals and university professors, GCAA's members are younger, more radical and with a background of student civic activism, as best exemplified in the figure of Tsui Shu-hsin, secretary-general of the association as well as a Taiwan National University graduate and documentary director.

8 During one interview, Linda Arrigo shared with me the following information, which she has in turn obtained through an interview with a former AEC official: allegedly, all throughout the 1980s, government workers received contributions to their welfare funds from scrap sales (Linda Arrigo, June 30, 2011).

9 Yenliao is the old name for the Gongliao area (Shih, 2012; 297); it is home to prehistoric relics from an aboriginal tribe, the Ketagalan Culture, as well as to a long-standing fishing community and several beautiful beaches such as Fulong, which hosts a variety of rock concerts and recreational activities all year long. The Yenliao Association is the most notable locally based anti-nuclear organization.

10 The operational nuclear facilities in the vicinity of the Taipei downtown area are respectively located at a distance of 28 kilometres (NPP-1); 22 kilometres (NPP-2) and 40 kilometres (the not yet operational NPP-4).

11 These anti-nuclear speeches and parades by intellectuals and cultural heavyweights are somewhat reminiscent of the early stages of the anti-nuclear movement in Taiwan, when scholars and intellectuals would travel to rural areas (下鄉) and impart their knowledge to villagers via speeches or articles (Ho, 2006: 43).

12 Green Citizen Action Alliance published its first anti-nuclear music album in 2013, which included 34 songs by local independent artists and bands, who play their music to promote anti-nuclear ideals among younger generations (Lee, 2014c: 3; Tsui, email communication, March 22, 2014).

13 During our interview, Professor Lin Tze-luen shared with me his views on how nuclear energy, always a political topic, has become even more politicized since the year 2000:

> If you are pro-nuclear energy, people think you certainly have to be a KMT supporter and if, on the other hand, you are opposed to it, you have to be pro-DPP. This of course is a biased and too simplistic view of the issue, as in Taiwan there are several individuals who do not vote for the KMT and nevertheless still think that Taiwan needs nuclear energy, in order to satisfy its energetic needs.
>
> (Lin Tze-luen, interview, July 15, 2011)

14 The *Liberty Times* survey question was phrased: DPP Legislators in the Legislative Yuan request a temporary halt of NPP-4 facility until a total check-up can be conducted, what is your opinion, in favour or against? [民進黨立委在立法院委員會以臨時提案要求核四停工進行總體檢，請問您是否贊成？]. In favor: 1180 (87.93%) [贊成: 1180 (87.93%)]; Not in favour: 160 (11.92%) [不贊成: 160 (11.92%)]; do not have an opinion: 2 (0.15%) [沒意見: 2 (0.15%)]. March 15.

15 The *Apple Daily* survey was phrased: Yesterday anti-nuclear groups established a meeting to organize an anti-nuclear parade, requesting for NPP-4's halting, what is your opinion? [昨反核團體串聯舉行反核遊行, 訴求停建核四, 請問你的看法?]. In favor of stopping NPP-4: 55.89% [應停建核四 55.89%]; not in favor of stopping construction: 31.54% [不應停建 31.54%]; do not have an opinion [12.57% 沒意見 12.57%] May 1, 2011.

16 Recently the KMT's stance seems to have changed once more and the Ministry of Economic Affairs has announced several times since 2013 that, in case controversial NPP-4 does not go into operation, then the life of all the other plants will have to be extended indefinitely. Considering that the normal retirement age for nuclear plants around the world is 28 years and that the Chinshan plant (NPP-1) already has 36 years of life, concerns about safety are increasing (Lee, 2014a: 1).

17 In 2013 the electricity generated by Taipower's 27 coal-fired power plants accounted for 69.4 percent of the country's total electricity production, while 11 hydroelectric plants generated 13.8 percent and three nuclear power plants generated 15.7 percent. As for the production of renewable energy, Taipower's 15 wind farms and three photovoltaic power plants accounted for less than 1 percent (Taiwan Insights, 2013). A report released by the government's organ Commission for Economic Planning and Development on August 12, 2013 listed the economic repercussions of stopping all nuclear operations. The report claimed that stopping the construction of NPP-4, while following the plan for the decommissioning of the three other plants, would cause the Gross Domestic Product to contract by NT$94 billion (US$3.1 billion) while 19,464 jobs would disappear thus leading to an increase in electricity prices, among other problems (*Taipei Times*, 2013a).

18 As of January 2013 this has become a Facebook page with 28,000 fans: www.facebook.com/pages/%E6%88%91%E6%98%AF%E4%BA%BA%E6%88%91%E5%8F%8D%E6%A0%B8/307969115955037?fref=ts.

19 Source: Jobin (2012): National percentage of votes for Green Party's electoral candidates in the 2012 elections, undivided per electoral districts: 1.7441%. Share of votes per candidate, divided per each electoral district: Taipei first electoral district (Zhang Yu-jing 張育憬) 2.34%; Taipei's seventh electoral district (Pan Han-sheng 潘翰聲) 24.00%; Taipei's eight electoral district (Yu Wan-ru 余宛如) 2.00%; New Taipei City first electoral district (Wang Chung-ming 王鐘銘) 2.37%; New Taipei City tenth electoral district (Yang Mu-wan 楊木萬) 4.58%; Taichung fourth electoral district (Cai Zhi-hao 蔡智豪) 2.00%; Changhua first electoral district (Shi Yue-ying 施月英) 1.52%; Tainan fifth electoral district (Lin Min-chang 林民昌) 1.52%; Kaohsiung third electoral district (Lin Zhen-Yang 林震洋) 1.66%; Taitung city (Xie Wen-Han 謝文漢) 0.72%.

20 A talked-about alternative solution could be to turn the plant into a natural gas-fired power plant, more environmentally friendly than coal-fired plants. Tsai Ying-wen's reasoning, which echoes the concern of many people inside Taiwan, starts from the assumption that the plant has already taken more than 14 years for its actual construction and has so far cost taxpayers about NT$300 billion (US$10 billion) (*Taipei Times*, 2013b).

21 In this regards, a German wind power company, InfraVest GmbH, has questioned the government's commitment towards increasing its quota of wind power and has accused the KMT of not being serious in its intention. In fact, after InfraVest announced that it was going to build 14 wind turbines on the pristine coastline of

Yuanli Township and some local residents, worried about the near distance between their homes and the turbines, launched several protests and hunger strikes, the government did nothing (Chen and Cole, 2013).

22 To be precise, only low-level wastes are stored in the deposit on Orchid Island; high-level radioactive wastes, requiring more than 10,000 years to decay, are stored in the cooling pools of the reactors, in the actual power plants. Should something happen to one of these plants, the radiation contamination could be made even worse due to the high amount of high-level radioactive wastes stored inside the plants, which could pose a serious hazard to the surrounding environment (Liu, 2011: 144–145; Huang *et al.* 2013: 1556). Furthermore, the six reactors' cooling ponds currently contain almost four times the amount originally planned, making Taiwan's power plants among the most dangerous in the world. According to Chiu Syh-tsong, head of AEC's fuel cycle department, the first fuel ponds will reach maximum capacity in 2014 (as quoted in de Changy, 2012). Concerning low-level radioactive waste, the repository on Orchid Island, which was supposed to be only a temporary home for such storage, has been full for the past five years, according to Professor Peter Wushou Chang, head of the Radiation Protection Association. His organization has detected that presence of caesium-137 in neighboring sweet potato fields and taro paddy fields, suggesting leakage (Peter Wushou Chang, informal communication, August 21, 2011).

23 In a survey of public perceptions of technological risks conducted by Chen Dung-sheng in 2009, 74.5 percent of respondents thought that nuclear wastes affected the health of humans, and 71 percent agreed that nuclear technologies pose unknown risks to both the environment and mankind (Chen, D., 2011: 570).

24 The becquerel (symbol Bq), named after Henri Bequerel who, together with Pierre and Marie Curie shared a Nobel prize for their work in discovering radioactivity, is the unit of radioactivity derived from the International System of Units (SI).

25 Considering that that instrument of the referendum had not yet obtained a legal status at the time, such ballots had no legal value; nevertheless anti-nuclear activists often quote those ballots, as the proof of popular opposition towards nuclear energy, in light of the fact that all of them generated an anti-nuclear majority (Ho, 2003: 699).

26 According to Taiwanese law, for a referendum to be declared "valid" the turnout of voters needs to be higher than 50 percent; furthermore, at least 50 percent of those who vote must say "yes" to the question, for the referendum to have legal validity.

27 Chu and Hau are among the most probable candidates to succeed Ma in the next presidential elections in 2016, thus indicating that their anti-nuclear stances were most probably calculated moves rather than bona fide anti-nuclear commitments.

28 Since none of the six earlier referendums held in Taiwan have managed to reach the threshold, protesters argue that the requirement for more than 50 percent of eligible voters to cast a ballot, for a referendum to be declared valid, is too high.

29 In regards to the occupation of Zhongxiao West Road, the Anti Nuclear Abolition Platform, which includes several anti-nuclear groups, announced on June 9, 2014, that many of its members have been requested to report to the nearest police station, for questioning in regards to their actions on April 27. Among them, Green Citizens' Action Alliance secretary-general Tsui Shu-hsin (崔愫欣) and Citizen of the Earth Taiwan's Taipei Office director Tsai Chung-yueh (蔡中岳) have both been told that they risk being charged with violating the *Assembly and Parade Act* (集會遊行法) and endangering public safety.

30 Lin Yi-hsiung, a former head of the DPP, is sadly famous inside Taiwan for the brutal murder of his mother and twin daughters in 1980.

31 While during the electoral campaigning for the presidential election in 2011 President Ma announced several times that the duration of the island's already functioning three nuclear facilities would not be prolonged, provided that NPP-4 be completed, the recent changes most probably signify that the other three plants will serve longer times.

4 Taiwan's petrochemical industry and the Kuokuang Petrochemical Naphtha Cracker

Background information

Mazu (媽祖) – goddess of the sea whose primary role is to protect fishermen and sailors – is of extreme importance in countries that have strong seafaring traditions such as Taiwan. According to the historical records, already in the seventeenth century numerous Fujianese left the mainland and came to Taiwan where they led lives whose habits were closely tied to the sea (討海 *tao hai*), on which they depended for both their livelihood and business traditions (Lin, L.H., 2010: 19). While many think about Taiwan as a modern – sleek and chic – country filled with skyscrapers and high-tech industries, the island presents a less well

Figure 4.1 Women cleaning oysters in Changhua, August 20, 2011 (source: Simona Grano).

known but extremely fascinating and culturally rich, rural side. In Taiwan's central area, set in Changhua County, we find one of the world's most amazing wetlands, populated by the endangered white sea dolphin (白海豚 *baihaitun*) and home to Taiwan's thriving oyster fisheries.[1] Even though the fishing industry is no match when compared to the financial revenues that a state monopoly such as the petrochemical industry can generate, it is nevertheless a thriving industry and, even more importantly, oyster fisheries are a precious cultural legacy representing Taiwan's rural origins. The proposed Kuokuang Petrochemical Project, planned for the wetland situated between Dacheng (大城) and Fangyuan (芳苑), on Taiwan's Eastern coast, would have presented a major threat to the economy and the ecosystem of the area, including its productive fishing industry, employing several thousand people. Further, Changhua is one of the island's main providers of rice, oysters, pork, eggs and vegetables.

The KPT project or eighth naphtha[2] cracker, was originally planned to be in Yunlin County (雲林縣 *Yunlin xian*) in the Litao Industrial Park (離島工業地區 *Lidao gongye diqu*) in 1995 (Ho and Shie, 2011: 38). Many years passed but the project remained a distant vision until a decade later, when talk about building it in Taixi village (台西), also in Yunlin, where 54 percent of locals were in favor of the project, resumed. In 2005, at the initial planning stage for the eighth petrochemical plant not many people cared about whether the project should go ahead or not and hardly anyone opposed it (Ho Ming-sho, interview, August 30, 2011). The only ones who were unhappy were local oyster farmers.

> In late 2007 and early 2008 … two major development projects, heavily polluting and with high-energy consumption, were planned at the site of the sixth naphtha cracker (六輕 *liu qing*) in Mailiao Township (麥寮), Yunlin County.
>
> (Dang, 2011: 8)

At that time Yunlin was still ruled by a KMT administration. Then, in 2006, a DPP legislator, Su Chih-fen (蘇治芬) was elected as county magistrate and the project became a political battle between the two main parties. After the DPP gained power over Yunlin, the environment towards said types of projects became less friendly and the investors realized that the project had lost some of its profitability and appeal. Further, due to the level of pollution release ascribed to the already up and running sixth cracker, the proposal for the eighth cracker failed to pass the Environmental Impact Assessment evaluation in 2008 and thus the investors decided to relocate and move to Changhua County, which was KMT-ruled, where local authorities had hinted that they would be welcoming to the investment. After the KMT won the presidential elections in 2008, KPT became a national investment project, deemed of major significance. In fairness though, the KPT project can be termed as a "bipartisan effort" since the plant was initially proposed during the DPP's national incumbency and was then supported by the KMT, which actively sought to make it operational (Ho, 2014b: 11).

The 2,773-hectare complex was set to be the second biggest on the island after Formosa Plastic Group's (台塑集團 *taisu jituan*) refinery complex in nearby Yunlin County, Mailiao (the already mentioned sixth cracker). The plant would have comprised several refineries, petrochemical plants, co-generation plants as well as an industrial harbor and was one of the key projects supported by the state, both for its annual production value as well as for its potential in creating new jobs in what is still an under-developed area (Williams and Chang, 2008: 178).[3] The project was estimated to require NT$620 billion (US$15 billion) in investment, a sum that makes NPP-4's investment pale in comparison (Ho, 2014b: 5).

The proposed location of the Kuokuang cracker was the 4,000-hectare stretch of wetland at the estuary of the Jhuoshuei River (濁水溪 *zhuoshui xi*),[4] which also forms part of the habitat of several wild bird species and of the white dolphin, whose population is believed to number fewer than a hundred nowadays.[5] While it is true that Taiwanese are increasingly aware of the need to protect animals, Taiwan is a recently developed economy where concepts such as "animal rights" and wildlife protection are still novel (Agoramoorthy, 2009), as seen in the numerous animals abused for tourist amusement. Activists say Taiwan's government does not actively enforce animal protection laws, responding only when there are complaints or media reports publicizing such issues.

Figure 4.2 Stickers and posters featuring the white sea dolphin (source: Simona Grano).

Figure 4.3 Sika deer as a tourist amusement in Green Island, August 15, 2011 (source: Simona Grano).

Figure 4.4 Sika deer in Green Island, August 15, 2011 (source: Simona Grano).

> Species such as butterflies and crabs receive protection mainly because they do not clash with economic growth.
>
> (Lai Fen-lan, interview, November 24, 2011)

In the KPT case, much to the white dolphin's detriment, the enemy was in fact a powerful one: the petrochemical industry.

A short history of Taiwan's petrochemical industry

Taiwan's petrochemical industry[6] has emerged in the 1960s as a consequence of state intervention (Chu, 2002: 3). Like other East Asian states, it started its first naphtha-cracking activities during the post-war industrial development in the late 1960s. After an upsurge of small enterprises dealing with downstream petrochemical industrial activities (especially the textile and rubber industries) took place in the 1950s, demand for mid-stream petrochemical raw materials grew exponentially, prompting the state to develop the petrochemical industry in Taiwan (Tu, 2001: 226). Afterwards, the government revealed its ambitious plan to produce upstream petrochemical material for internal consumption (Ho, 2014b: 5–6). At the initial phase the government was the sole investor in the petrochemical sector, which was entirely dominated by its state-owned firm China Petroleum Corporation (henceforth CPC). CPC built its first naphtha cracker plant in 1968, with an ethylene capacity of 54,000 metric tons per year (MTY). By 1984 it had four crackers with a combined ethylene capacity close to one million tons per year, ranking about 11th in the world. The Formosa Plastic Group, the largest private chemical firm in Taiwan since the 1960s, established with financial loans from the United States, requested to build a plant but was turned down by the government in 1973 (Ministry of Economic Affairs, 1999). In 1986, after persistent pressure from petrochemical private firms and their lobbying efforts, the government granted permission to Formosa Plastic to build Taiwan's first private cracker plant (Taiwan's sixth) while at the same time allowing state-owned CPC to go ahead with the construction of a long-delayed fifth cracker (Chu, 2002: 6). After 1986, the government closed the first and second crackers in response to local environmental protests, which also caused delay in the construction of the fifth[7] and sixth crackers. As a result the fifth cracker did not begin operations until 1994, while the sixth one was not up and running until 1999 (PNN, 2010). While the sixth naphtha cracker was initially planned to be set in Ilan, strong opposition pushed FPG to eventually relocate the project to Yunlin County in 1991 where, differently than in Ilan, political opposition (DPP) was not strong and job opportunities scarce (Williams and Chang, 2008: 77, Jhan, 2011: 51).

After that, two more plants were proposed: Tuntex's seventh naphtha cracker and CPC's eighth naphtha cracker plant. While the seventh plant was never built, the eighth one was planned by CPC Corporation with the aim of relocating its crude refining plants in southern Taiwan's Kaohsiung City to Changhua County by 2015 (Ho, 2014b: 11).

As stated by Ho, Taiwan is not an ideal location for petrochemical plants, due to its lack of natural gas and petroleum (Ho, 2014b: 5). Such natural deficiencies, however, have not prevented Taiwan from turning into what some economists have termed a "petrochemical kingdom" (石化王國 *shihua wangguo*). In fact, Taiwan's petrochemical industry has been undergoing a constant upgrading of its capacity and is geographically spread out over the entire territory of the island, thus making environmental – and related health issues – quite evenly distributed around the country; for all of these reasons, the petrochemical industry is best suited to convey an in-depth glance at the changing contours of environmental politics in contemporary Taiwan.

Main problems pertaining to the Kuokuang Petrochemical Project

The major concern regarding the KPT project was the impact that such a facility would have on the local wildlife and ecosystem and the negative effects – in terms of damages and losses – for the local fishermen and oyster-farmers community.[8] The mega-project involved large-scale land reclamation as well as the construction of harbors and industrial facilities, which would have wreaked havoc on the local ecosystem.

In the past decades numerous wildlife species have transformed the Dacheng wetland into a highly diversified and rich ecological system, the habitat of several birds and crabs species, some of which are eligible for protection under the *Wildlife Conservation Law*.

Second, Changhua County's soil is amongst Taiwan's most fertile. Unfortunately, due to extensive construction of industrial complexes – with their thirst for water – and underground water-pumping activities operated by local farmers, the area has become one of the most affected by land subsidence problems.[9]

Renowned farmer/writer Wu Sheng, whom I have had the pleasure of meeting on several occasions, once told me:

> We just want to have enough water so we can keep farming all the while protecting Changhua from pollution, one of Taiwan's most important farming areas. In the past, water used to rise to 166 inches (422 centimetres), now it gets only 100 cm high making the time to get oysters much longer, due to the nearby industries all pumping water away from under the earth.
>
> (Wu Sheng, interview, October 16, 2011)

As he told me when we met in Erlin (二林) during a protest against the fourth expansion phase of Taiwan's Central Science Park (中科四期 *zhongke siqi*):

> It is not simply about saving the endangered white dolphin, the KPT will also destroy Taiwan's farming and fishing culture that has been passed down for generations both by making it harder to farm oysters and by polluting the water where they grow.
>
> (Wu Sheng, interview, October 15, 2011)

Figure 4.5 The author with writer/poet Wu Sheng, in Changhua County, Erlin Township, October 15, 2011 (source: Simona Grano).

Changhua's problems of land subsidence and salinization date back to the 1970s. Prior to this decade these lands were mostly rice paddies. In the 1970s and 1980s Taiwan underwent a period in which rice production was in excess and thus the government started to encourage diversification. Consequently, many of these rice fields were converted into artificial fish ponds for raising eels, shrimps or oysters (Williams, 1992: 192). With their massive demands for water, such ponds induced saltwater to seep in underground aquifers, causing intense land subsidence. Consumption of groundwater, especially in the central and southern coastal areas, averages more than 7 billion tons a year (Williams and Chang, 2008: 109), resulting in a chronic deficit of water. Second only to Pingtung County, Changhua has the most severe land subsidence problem in all of Taiwan. In the most affected areas around Fangyuan and Dacheng, land sinking (土地的下陷 *tudi de xiaxian*), resulting from excessive groundwater utilization, can be as deep as an entire floor (Lin, R., 2010: 2). Land subsidence is also spreading farther inland. Due to the sinking of the coastal area, which has been taking place for decades, seeped-in seawater has caused extracted groundwater to be salty and the soil to become salinized and worthless in terms of agricultural productivity. In Dacheng, where the economy used to rely heavily on agriculture, the acreage of abandoned farmland increases every year.

Animal lovers, environmentalists and concerned scholars believed that another huge petrochemical complex would have harmed the natural environment, endangering people's health as well as their traditional ways of life while further aggravating the area's land subsidence problems (Ho Ming-Sho, interview, August 30, 2011; Wu Sheng, interview, October 15, 2011; Lai Fen-lan, interview, October 30, 2011).

According to Dr. Lee Hong Yuan (as quoted in Lin, R., 2010), Professor at the Department of Civil Engineering at National Taiwan University, CPC, KPT's developer and the local government failed to address four main key points regarding the petrochemical facility:

1 Given that the complex should have been located in a land subsidence area already lacking enough water for local farmers' consumption, how and from where was the plant going to secure the amount of water it would need to function?
2 If land subsides even further, which measures would the developer/local government have applied in order to prevent flooding?
3 What could be done to solve the problem of the increasing salinization of the soil?
4 And what is to be done about worsening erosion caused by seawater?

According to Dr. Lee the government acts irrationally in regards to industrial development, water resources management and land use and its policies regarding coastal protection and agriculture lack any serious long-term consideration (Lin, R., 2010). To support his claims he calculated that Changhua County has 1.31 million inhabitants who consume *c.*380,000 tons of water every day. The KPT project would have needed 400,000 tons per day; furthermore the Erlin branch of Taiwan's Central Science Park, which has already passed the Environmental Impact Assessment, will also be located in this water-thirsty county.[10]

Figure 4.6 Protest against the fourth expansion phase of Taiwan's Central Science Park and its water consumption, in Erlin Township, October 16, 2011 (source: Simona Grano).

Figure 4.7 Protest against the fourth expansion phase of Taiwan's Central Science Park and its water consumption, in Erlin Township, October 16, 2011 (source: Simona Grano).

Even though its agrarian productivity has declined in the past decade, Changhua is still one of the island's most important agricultural sites, with numerous daily-consumption items from rice, eggs, pork meat, vegetables and fruits being produced here, thus leaving people concerned about the safety of food products derived from areas set in the vicinity of a petrochemical complex.

Such anxieties about the quality of Taiwan's own domestic products were further exacerbated by the already mentioned food-scare that exploded in May 2011, after Taiwan's Ministry of Health announced that they had discovered that food-addictive supplier Yu Shen Chemical Company had illegally added carcenogenic plasticizer DEHP in clouding agents and sold toxic agents to several food and beverage producers around the country. Toxic substances were found in five categories of daily consumption goods: sports drinks, juices, tea drinks, fruit jams and fruit syrups.[11] Further concerns and anger exploded later on that summer, as it was discovered that DEHP contamination was not limited to such products but had been found in pharmaceutical products (vitamin pills and supplements) and baked goods such as bread, in July 2011 (*The Economist*, 2011).[12]

Due to the fact that petrochemical plants are to be found all over Taiwan's territory, previous examples of how such complexes harm the soil and water resources nearby are plentiful (e.g., the damages caused by the Mailiao plant are by now well known) and serve as a constant reminder to villagers of a fate they want to avoid in their own localities; an issue that will be dealt with later on in the course of this chapter.

Professor Wu, a chemical engineer and one of the few Taiwanese experts of health risks assessment, had doubts that the government was an impartial stakeholder in the KPT project and that the EIA process had been handled in a transparent way. "An example of how ridiculous the EIA assessment can turn out to be is to be found in the EIA regarding CPC's plant in Kaohsiung, where wastewaters were determined to be cleaner than drinking water!" (Wu Kuen-yuh, interview, September 20, 2011). Additional problems that contributed to fuel opposition to the project, were the potential health damages – in terms of an increase in the cancer rate in the Changhua area due to toxic gases inhalation and polluted-water discharges – for local citizens. Again, in Professor Wu's words:

> A colleague of mine, Jhan Jhang-cyuan (詹長權), conducted an epidemic study using the statistics of the National Health Department and found out that cancer occurrence and mortality rates in two nearby townships (Mailiao and Taixi) had increased since the sixth naphtha cracker started its operations.[13] If we now locate yet another cracker in central Taiwan, you can imagine what the outcome will be.
>
> (Wu Kuen-yuh, interview, September 20, 2011)

Some experts go even further in their critique by claiming that the proposed eighth naphtha cracker, as well as the operational sixth naphtha cracker in Mailiao, run counter to the government's policies on energy-saving and reducing carbon dioxide emissions (Loa and Shan, 2010).[14]

> Kuokuang Petrochemical would have been responsible for 25 percent of the nation's carbon dioxide emissions by itself, when operational. Furthermore, by now the negative effects of the sixth cracker on land, air and water resources in areas near the plant, are well known.
>
> (Chou Chang-hung, interview, June 30, 2011)

The anti-Kuokuang movement

Those who supported the KPT project claimed that a 2,000-hectare plant, which was to include a 300,000-barrel a day refinery with factories producing 25 different types of chemical products, would have significantly increased Taiwan's capacity to produce ethylene (Sui, 2011b).[15] Since ethylene is used in construction and plastic and it is very much requested by China, they assert that the KPT project would have benefited Taiwan's coffers (Chou Kuei-tien, interview, October 14, 2011). The Ministry of Economic Affairs further estimated that the project could have attracted NT$933.6 billion (US$29 billion) in future investments, while generating NT$460 billion in annual output and creating 375,000 new job opportunities (Wang, 2010). According to one of my interviewees, the already-mentioned writer Wu Sheng, behind the scenes were the private incentives of local officials and their close-knit ties to the developers (Wu Sheng, interview, September 16, 2011).

Wu Sheng, a native of Changhua, is actually one of the leading figures in the anti-Kuokuang movement. A retired junior high-school biology teacher as well as a self-taught poet/writer he has composed several poems and written articles on the area's natural beauty and on the effects of pollution (Wu, 2011: 4); some of these have been edited, together with essays and poems from other authors, in a collection titled *Wetland, Petrochemical Industry and Island's Visualization* (溼地, 石化, 島嶼的想像 *shidi, shihua, daoyu de xiangxiang*), co-edited by himself and writer Wu Ming-yi (吳明益). This book represents an active attempt at trying to influence Taiwanese' feelings towards their country's natural environment. Portraying several different points of view, from that of literary scholars, to journalists, to chemical and toxicology experts such as Jhan Jhang-cyuan, to local residents and artists, it tries to convey how the petrochemical industry would impact on locals, their livelihood and on the surrounding natural habitat, by destroying water resources and agricultural and fishery products; most of all, in light of the damages provoked by the already running sixth plant, the book emphasizes the negative impacts the plant would have on local residents' health. Being extremely well known both inside Taiwan as well as abroad (he was invited to the Iowa Writers' Workshop as a visiting guest as far back as the 1980s (Wu Sheng, interview, 16 October, 2011) Wu Sheng's words have a deep resonance among the general public.

In the anti-Kuokuang movement, the involvement of "celebrities" from the artistic and cultural community and their high status in society has no doubt contributed to give a great boost to the anti-petrochemical movement with film directors, teachers and poets joining more traditional stakeholders (e.g., activists, journalists, politicians, local residents, scholars) and symbolizing, in my opinion, one of the most interesting new additions to the environmental governance process of the country.

Countless campaigns and petitions to fight KPT were organized by the medical community, by university professors and *Academia Sinica* scholars aided by the cultural circles of Taiwan.

Other – heavily involved – groups and stakeholders in the anti-Kuokuang movement were local farmers and oyster fishermen together with students, medical experts, lawyers and scholars from all over the country as well as a few religious groups. It has to be said that the anti-Kuokuang campaign enjoyed a very limited support among residents of Dacheng. In Fangyuan, however, a

Figure 4.8 Writer Wu Sheng with writer/film director Hsiao Yeh (小野) in Changhua, October 15, 2011 (source: Simona Grano).

famous oyster production site with past environmental protest experience, residents readily joined the opposition. Furthermore, several important national NGOs supported local activists in their anti-Kuokuang struggle such as the Taiwan Society of Wilderness (荒野保護協會 *huangye baohu xiehui*), the Wild at Heart Legal Defense Association (臺灣蠻野心足生態協會 *Taiwan Manye Xinzu Shengtai Xiehui*), the Green Party of Taiwan (綠黨 *lüdang*), Taiwan Rural Front (臺灣農村陣線 *taiwan nongcun zhenxian*), the National Youth Alliance Against Kuo Kuang Petrochemical Project (全國青年反國光石化聯盟 *quanguo qingniang fan guoguang shihua lianmeng*), The Alliance for the Protection of the Pink Dolphin (臺灣媽祖魚保育聯盟 *taiwan mazuyu baoyu lianmeng*), Taiwan Environmental Information Association (*taiwan huanjing zixun xuehui* 臺灣環境資訊協會) and the Youth Synergy Foundation (青平台基金會*qing pingtai jijinhui*), turning what was initially a local issue into a campaign of national concern and visibility. They attacked the project on multiple fronts: its impact on the environment, its impact on the local economy (e.g., oyster farmers' losses), its impact on residents' health and whether the petrochemical industry is positive or not for Taiwan's future development.

In a down-to-earth publication activists from Taiwan Rural Front employ a powerful narrative to illustrate to the public the health hazards predicted by toxicology experts; they do so in a simple way, in order to reach the widest possible audience, including those who would not be able to understand complex scientific explanations (Taiwan Rural Front, 2010). However, this book is more than a simple vulgarization of scientific data; through the collection of oral histories of Dacheng residents, coupled with comparisons with international standards on the treatment of petrochemical wastes, they sketch a complex trajectory of pollutants and their impacts on people, fisheries, oysters and local the ecosystem.

Students and younger activists employed newer and more creative tools to express their dissent by altering scenes from the movie *Avatar*, by James Cameron, and juxtaposing these images with pictures from the wetland's natural habitat, trying to convey the implicit message that the Taiwanese government was mistreating its citizens in the same way the company in the film exploits Pandora's natives, all the while highlighting the negative impact the plant would have had on the local flora and fauna.[16] Such graphic representations of environmental destruction hark back to Taiwan's democratization struggle during the 1980s, when the politicization of local Taiwanese identity versus an evil and voracious ruler that had come from outside, played a crucial role.

Thanks to the weight of such concerted efforts, by the summer of 2010 several important media outlets had run stories on the anti-Kuokuang movement. In that same year activists launched an environmental "trust-fund" (信託法制 *xintuo fazhi*) campaign to save the land (Lu Shi-wei, interview, September 14, 2011). The goal was to induce the public to donate a symbolic amount of money to "adopt" a parcel of the coastal wetland that the developer, CPC, wanted for its petrochemical plant. In the end more than 56,000 people pledged *c.*NT$8 million (Sui, 2011a), enough to purchase a little over 200 hectares of coastal wetlands in Changhua, along which lives the endangered Indo-Pacific humpbacked dolphin, the scientific name of the already mentioned white sea dolphin. According to Tsai Chia-yang (蔡嘉陽), Changhua Division chairman of the Taiwan Environmental Protection Union[17] after the first stage of the land trust, they planned to collect more money and purchase another 800 hectares of wetland, which constitute an essential habitat for some rare bird species. Each piece of land (1 m^2) was priced at the symbolic sum of NT$119 (119 being the phone number for emergency services in Taiwan) (Loa, 2010b). Nevertheless, the former Deputy Minister of the Interior Lin Tsyr-ling (林慈玲) released a statement saying that in case green groups could not obtain permission from the National Property Administration to purchase the land, the Ministry could not allow them to set up a trust fund (Loa, 2010a). In the end the campaign could not get started because of this "legal obstacle" (Lu Shi-wei, email communication, September 16, 2011).

After Wu Den-yih's (吳敦義)[18] controversial outing, in which he claimed that white dolphins should not be a matter of concern since they can actually co-habit with petrochemical projects by "swimming around them" (as quoted in Chen, Z.L., 2011: 91), young activists, among them singer Wu Zhi-ning, Wu Sheng's son, composed a song titled *The Song of the White Dolphin* (白海豚之歌),[19] turning the conservation of this quasi-extinct mammal into one of the key assets of the protest. The dolphins thus became the symbol of the cruel fate awaiting all living creatures inhabiting the wetland, namely the destruction of their natural habitat, encroached by pollution, noise, wastes and poisonous emissions, released by petrochemical plants.

In the light of all the tactics expounded by activists, it becomes evident that what Weller said about Taiwan's environmentalism being constituted by a mix of different elements, from international practices to partly native elements, an "alternative civility" where environmentalists mix and match different methods and ideas to spread ecological awareness, reflects a still up-to-date version of Taiwan's environmental activism. The trust fund campaign, for instance, drew its inspiration from previous mobilization activities by international organizations such as the World Wildlife Fund (WWF) and Greenpeace; such globally oriented tactics are mingled with native cultural symbols like the employment of the number 119 and the protection of the island's own distinctive, maritime cultural legacy when it comes to oyster fisheries and the frequent recurrence of the Goddess Mazu. Again, folk religion and native cults get re-adapted and intermingled with Western-imported ideas, tailored to local specificities. The organization of seemingly foreign elements such as the use of rock concerts to express dissent, are localized and indigenized through the single-handed use of *Taiyu* as the lingua franca of activists, thus consciously identifying the native Taiwanese element with the "good" and the "positive" (the protection of nature), while almost completely ignoring the Chinese element; in fact, all the songs as well as the poems composed by activists/artists rely predominantly on Taiwanese or on Mandarin forms heavily influenced by Taiwanese.

Just like in the fourth nuclear power plant's case, the political element cannot be underestimated. In fact, while we have seen that this project was, in its initial planning phases, supported by both parties, once the anti-Kuokuang movement became heavily featured on mainstream media both the DPP as well as the KMT underwent a complete change of direction, to please the public. Wen Chin-chu (翁金珠), a local DPP county magistrate who had attempted to develop the Dacheng wetland while in office (2002–2005), joined the opposition as soon as public skepticism started to rise in 2010. Her changing position, from environmental protection safeguarder to supporter of local development and back, came to be seen as a betrayal of her original environmental commitment during the early stage of her career, and was considered as a clear indication of how the DPP can easily switch its position for electoral gains. When I talked to poet/writer Wu Sheng, he repeatedly stressed – and made sure that I would truly comprehend – how much the Kuokuang case, as well as many others for that matter, was enmeshed and deeply entrenched in local and national political interests (Wu Sheng, interview, October 15, 2011).[20] In 2011, Tsai Ying-wen and Su Tseng-chang, national DPP candidates and contenders in the party's primary for the soon-to-be presidential election, openly declared their support of the anti-Kuokuang movement. Just like county magistrate Wen, their jumping on the opposition bandwagon was seen as a sudden turn-about due to the looming elections. In fact, as mentioned before, the Kuokuang project was actually first proposed in 2006, when the DPP was in power with Su and Tsai serving respectively as premier and vice-premier at the time (Ho, 2014b: 12). Politicians on all sides have their careers in mind and only seem to care and cater to popular wishes when public discontent rises, especially in a pre-electoral year.

An element that has contributed to render this protest different than those of the past is the active and vocal participation of hundreds of middle and high school students in the fight against the petrochemical project, mainly in Changhua County but also in Taipei, where they took part in numerous night watches in front of the Environmental Protection Agency. Some spontaneously joined the protests while others were brought to Changhua in order to take part in a series of educational activities organized by social groups, regarding the life of Taiwanese farmers and fishermen, in order to pass down to younger generations the knowledge in regards to the island's traditional way of life and to show students what these people would be losing if the plant were built (Tsai Pei-hui, informal communication, September 22, 2011).

Many are of the opinion that it is precisely this heart-felt participation, considering the pressure faced by Ma Ying-jeou, hoping to be re-elected the following year, which more than anything convinced Ma to stop the petrochemical project on April 22, 2011, because it had become a liability (Tang, 2011: 3).

Framing the protest: making use of past experiences

Between 2010 and 2011 a variety of civil groups joined in the collective effort to stop the eighth naphtha project. While some were in it for private interests – such as the associations fighting for higher compensation on behalf of fishermen – most had more altruistic motivations, dealing with the preservation of rare species of birds and other coastal creatures. As seen in the previous section, green activists held numerous protests all over Taiwan from Taipei City to Changhua County. The main protest took place on November 13, 2010 as nearly ten thousand people gathered in front of the Presidential Palace, on Ketagalan Avenue, to complain against the government-backed plan of building KPT in Changhua.

A frequently analyzed factor, influencing the relationship between protesters and pollution is that of "framing" (Benford and Snow, 2000). Various articles on framing illustrate how the same phenomenon can be viewed and perceived differently by different sets of people (Vaughan and Seifert, 1992). In the KPT's case protesters framed the environmental threat in a way that echoed Fangyuan villagers' cultural values and orientations about lineage and maritime traditions. The emphasis was on the potential destruction of their lineage, due to the harmful effect of pollution and undeniable destruction of their heritage as oyster farmers, passed down from generations. First, enlisting the help of many famous artists, writers, film directors and singers, locals created several fake newspaper articles depicting the frightening scenario in the town of Mailiao, at the site of the sixth naphtha cracker and the destruction provoked by the petrochemical industry. Following the setting of the petrochemical complex in Yunlin, the environment has become seriously degraded and although it is not easy to establish a direct link between pollution and health damages (Peter Wushou Chang, informal communication, August 31, 2011; Wu Kuen-yuh, interview, September 20, 2011), it is undeniable that the complex has generated a visible amount of water and air pollution (Jhan, 2011: 48–54). In Yunlin agricultural repercussions and backlashes played an important role in the increase of social discontent. In fact, numerous agricultural products saw a drop in sales after the sixth plant became operational, for fear of poisoning. This upset many villagers who had long thought of their hometown as a place of natural beauty and healthy agricultural products. Locals living off the land had not only lost their source of income from planting crops but also had to spend more money to buy their own food. In 2012 Professor Jhan Jhang-cyuan released the results of a three-year study, sponsored by the local Yunlin government, which demonstrated the increase in the rate of cancer in an area set within ten kilometers of the sixth naphtha cracker. With such a precedent in mind, Changhua oyster and cash crop farmers had even higher incentives to oppose KPT. Further, villagers and fishermen of Changhua understood the potential adverse effects of chemical pollution thanks to earlier industrial incidents. In fact, while during normal production residents living close to a polluting factory may have difficulties linking health disturbances to the pollution's release, in case of an accident health consequences are immediately visible. Several explosions at the site of the sixth naphtha cracker releasing toxic gas, made it almost impossible for locals to breathe (Taiwan Environmental Information Center, 2010). Further, Professor Tsuang Ben-jei (莊秉潔), an epidemiologist at National Chung Hsing University, has carried out a study that pinpoints the damages caused by the sixth naphtha cracker, mostly dealing with local residents' health, in terms of an increase of cardiovascular diseases and cancers. In 2010 Professor Tsuang created a computer model of the aereodynamic flow of

Figure 4.9 Fake newspaper publicizing the defeat of the petrochemical industry after KPT's cancelation (source: Simona Grano).

fine particulate matter smaller than 2.5 micrometers (PM2.5) if Kuokuang ever became operational and discovered that the plant would shorten the life expectancy of locals on average by 23 days (Prof. Tsuang, email communication, November 12, 2012). His research has won him some powerful enemies among the petrochemical sector – and particularly within the Formosa Plastic Group – developer of the sixth naphtha cracker, which is currently suing Professor Tsuang for "harming the company's reputation" (Lu, 2013). Attorney Chan, of Primordial Law Firm – one of my interviewees – is part of the legal team that defends Tsuang in court.

It is worth noting that, among the several reasons that make this case special, activists, aided by several independent professors and experts, among whom are Professor Tsuang Ben-jei, Professor Jhan Jhang-cyuan, Professor Wu Kuen-yuh, Professor Chou Chang-hung, Professor Chen Chi-chung (an economist whose role will be analyzed in depth in the next section), and a few others for the first time carried out an independent estimation of the costs, which was presented to the EIA Committee during the EIA process and served, among the evidence collected by the anti-Kuokuang faction, as tangible proof that the project should be dropped because it is not cost-effective. The next section deals with the legal tools and participatory approaches employed by the opposition movement to fight this project within the EIA system.[21]

Legal aspects and rules-based participatory approaches

Although this case is considered as a milestone since it has ignited the strongest opposition to an industrial project in Taiwan in decades (and as such has been the focus of various studies and books inside Taiwan, e.g., Taiwan Rural Front, 2010; Ho and Shieh, 2011; Wu and Wu, 2011), a barely researched aspect is the utilization of those legal channels, available due to the introduction of participatory channels imported from democratic systems, which activists have sought to employ in order to push forward their agenda, and will be at the center of the present section.

In one of the numerous interviews I have conducted on the KPT case, I was lucky to be able to talk with a young lawyer activist by the name of Lu Shi-wei (陸詩薇) who works for the Taipei-based environmental law association, Wild At Heart.

In the KPT case she acted as the lawyer of the students involved in the protest, which, as seen in the previous section, was characterized by a high level of participation among younger generations and students all over Taiwan Island.

Lawyer Lu told me that:

> From a legal point of view, the state had no choice but to fight (爭開發 *zheng kaifa*) the project because it ran against the *National Preservation Project of Taiwan's Shore and Tidal Lands*, which is a project set up by the R.O.C. government in 1973 with the aim of protecting Taiwan's shores.
>
> (Lu Shi-wei, interview, September 14, 2011)

The first legal breach was the geographical location of the plant, which, according to Lu Shi-wei, was illegal. In fact, according to article 5, subparagraph 2 of the *Standards for Evaluating the Environmental Impact of Developmental Activities*:

> if a developing activity is located in an area which, according to another law [in this case the *National Preservation Project of Taiwan's Shore and Tidal Lands*] is not allowed to be used, then, the EIA Committee has no choice but to reject the developmental project (不通過的結論 *bu tongguo de jielun*).
>
> (Lu Shi-wei, interview, September 14, 2011)

In other words, the dilemma the protesters had in front of them, as they enlisted Wild At Heart's help, was to define whether Dacheng could be considered as a legally prohibited area or not.

According to the *National Preservation Project*, Dacheng falls in the category of a medium range protection zone (一般保護區 *yiban baohuqu).* One of the basic regulations in place to protect such areas, requires that land and sand should not to be taken freely and that any project planned in these areas shall "blend in" (融入 *rongru*) with the natural surroundings, being barely noticeable. In this regard, protesters and opponents pointed out that the KPT complex would have needed huge amounts of sand and rocks, which would have had to be extracted from somewhere and, second, in Lu's words: "It is unimaginable that such a huge cement monster will be able to *blend in* with the natural surroundings, being barely noticeable" (Lu Shi-wei, interview, September 19, 2011). Therefore, from a legal perspective, the choice of building KPT on the wetland between Dacheng and Fangyuan ran against the already settled principles of the above-mentioned regulations, established by the central government.

Local villagers and residents believed that pollution from the complex would have harmed their health, the ecosystem and would have had life-threatening consequences for species such as the endangered white dolphin; however, they understood that to stop the complex they needed more. Based on this understanding, instead of framing their protest in a pure environmental perspective, together with a group of lawyers from Taipei, they framed their claims mostly in terms of "violation of land-related regulations." This piggybacking on land-related issues, rather than concentrating solely on ecological concerns, in order to give their claims more strength, proved a successful strategy to avert a possible environmental catastrophe.

Specifically, the land where the petrochemical plant was planned belongs to the ROC government but since KPT was deemed as an "extremely important project for Taiwan" by the Legislative Yuan itself, there were no reasons for the government not to lease or sell the land to CPC. According to the laws regulating the use of non-urban land (*Non-Urban Land Use Regulations*) if a developer does not have a special permission (which CPC did not have) it cannot locate any developmental activity within 3 km from a wetland (subparagraph 9) signifying that CPC was breaking the law of the ROC in regards to coastal areas (Chan Shun-kuei, interview, November 18, 2011).[22]

Apart from carrying out EIAs for controversial or massive developmental projects such as petrochemical plants, Taiwan has a list of development-related activities that should also undergo an EIA process due to the potential damages they might cause to the natural surroundings. For example, the act of taking rocks and stones (the raw material), as the CPC planned to do in order to build the petrochemical complex, is a massive process that can provoke incommensurable damages to the ecological environment of a place. When I interviewed Professor Chou Chang-hung (周昌弘), he asked me a simple yet exemplifying question:

> Where would CPC take the 288 million rocks they have estimated they would need? They have never explained it therefore I believe that for the Committee responsible for the EIA process it would be impossible to decide whether to approve KPT or not since they would have to first carry out an additional EIA simply to estimate if the recovery of sufficient rocks would be feasible or not and from where these would come from.
>
> (Chou Chang-hung, interview, June 30, 2011)

Since this additional EIA has not been carried out, Lu and her colleagues from Wild at Heart have not included this issue in the proceedings. A document submitted to the court by the defendants and their lawyers, in order to set an obstacle for CPC, argued that first EPA should carry out an EIA to assess the problems caused by the act of collecting the raw material needed for the construction, and only secondarily one on the KPT Project itself.[23] Furthermore, the above-mentioned *Standards for Evaluating the Environmental Impact of Developmental Activities* state that if waste water from one of these developmental projects is going to be discharged in a river or in some other water body, the firm responsible for the project (in this case CPC) first has to carry out an evaluation of this impact, which, according to my informants, in the KPT case was never carried out. Therefore, Prof. Chen Chi-chung (陳吉仲) is of the opinion that to be able to fully gauge the total costs that KPT could have generated versus its profit, the state should have first carried out an EIA taking into consideration the impact on the livelihoods of the fishermen affected (calculating the amount of money for their compensation) and also of rivers and oceans ecology after the discharge of wastewaters (Professor Chen Chi-chung, email communication, August 31, 2011). For this and other reasons, Prof. Chen (an economist who specializes in cost-benefits analysis) and Professor Wu (the already-mentioned toxicology expert previously involved in the health-damages analysis in the Mailiao cracker case), estimated that, due to the amount of additional expenses not taken into account by CPC (for fear of driving up its costs) derived from the potential increase in cardiovascular and respiratory diseases, the rising unemployment rates among fishermen, the impact on tourism and seafood industries, the increase in greenhouse gases' discharge emissions (and subsequent potential sanctions by the international community for failing to meet the standards) the total cost of the KPT project would have fluctuated from NT$56.9 billion to NT$112.1 billion per year; as a consequence, the financial returns would be considerably lower for Taiwan by a total of NT$36.5 billion per year (Chen Chi-chung, email communication, August 31, 2011)![24]

In Chen Chi-chung's words: "According to the cost-benefit analysis of Kuokuang Petrochemical Project, the total cost is higher than the benefit. In addition, this project will lead to social unfairness. Therefore, it should not be executed" (Chen Chi-chung, email communication, August 25, 2011). According to Lu Shi-wei, an EIA on how to compensate oyster farmers (or, for that matter, on how to quantify the damage they would suffer) was never carried out. CPC did in fact produce an EIA but, according to green activists, it was a laughable "façade" document – half a page long – where CPC stated its plan to set up an "ecology and environmental-health insurance fund" without specifying its content, what was covered in it and how the money raised in the fund would be spent (Lu Shi-wei, interview, September 19, 2011).

This case illustrates how participatory approaches for public participation in environmental planning in Taiwan (with the EIA as the most significant participatory channel to date) are still far from perfect. Professor Wu Kuen-yuh is convinced that the EIA conducted for the KPT project was tilted towards the interests of the business community. He furthermore claims that the health and environmental risks connected to the petrochemical industry were already known ten years ago but that the official reports, which he had read, were suppressed by the government (Wu Kuen-yuh, interview, September 20, 2011).

Even though in the case of KPT, governmental industrial policies have been influenced by several different agents, who attempted to sway the final outcome by employing those available participatory approaches, the final decision regarding the cancelation of the petrochemical plant was directly linked to political interests. The next section will deal with such political interferences and fallouts.

Political influences and strategic timing

While doing research on this case study, a thing that never failed to surprise me was that among the many activists, experts and scholars interviewed, the predominant opinion was not to consider this case as a triumph (Robin Winkler, interview, September 5, 2011; Lu Shi-wei, interview, September 16 and 19, 2011; Tsai Pei-hui interview, September 22, 2011; Lin Le-hsin, interview, September 24, 2011; Ho Ming-sho, informal communication, November 24, 2011).

According to Tsai, spokesperson of Taiwan Rural Front: "The environmental impact assessment was dominated by political forces" (Tsai Pei-hui interview, September 22, 2011). First of all activists reject the procedural process in which they obtained the halting of the project arguing that the EIA committee, in light of all the legal documents provided, should have immediately thrown out the case; instead, on April 22, 2011 it put forward two available choices: (1) stopping the project; (2) going ahead with the project upon approval of certain conditions. Activists lament that procedurally this case should've been met with refusal on part of all the members of the Environmental Impact Assessment Board (環評大會 *huanping dahui*); instead, it was President Ma Ying-jeou's political statement that settled the case on that same day. As a result the developer withdrew the project. The *Liberty Times* speculated in one of its editorials that it was purely out of political concerns that Ma decided to cancel the plant (*Liberty Times*, 2011). In other words environmentalists are convinced that this course of action has weakened the institution of the EIA consequently failing to establish a legal precedent, which would have been useful for future references. According to Prof. Tsai this solution left the impression that the case was given a green light by the EIA Committee and was then subsequently killed by politicians. "It doesn't bode well for the future" (Tsai Pei-hui, interview, November 3, 2011). Soon after the decision, an additional motive of distress was that, when the KPT project was "killed," President Ma merely said that the plant wouldn't be built in Changhua; he didn't rule out its reincarnation elsewhere (Ho Ming-Sho, interview, August 30, 2011).[25]

After the KPT project was officially canceled several local politicians, including Cho po-yuan (卓伯源), county magistrate of Changhua since 2005 and a KMT affiliate, harshly criticized President Ma for being too weak and for kowtowing to environmentalists. Local politics, especially in impoverished areas, are among the most important factors in steering through developmental projects, which have the potential to deliver much-needed financial revenues.

The stopping of the Kuokuang Petrochemical Project is often heralded as a major success for environmental groups, green academicians, artists, writers and concerned citizens. When asked why she thinks the KPT project was eventually canceled, lawyer Lu told me that it reached a halting point because all powers in society, from scholars, to lawyers, to students to the academic community staging sit-ins and protests in front of EPA were a powerful force pressuring President Ma to reconsider the plans, out of fear of losing the upcoming presidential elections. In fact the movement generated a great media exposure, which due to the approaching elections, put the government under significant pressure. Thus, strategic timing in taking advantage of such a crucial moment on the part of protesters was fundamental for the "positive" outcome of the anti-Kuokuang campaign.

Some DPP candidates utilized the anti-Kuokuang stance as a golden opportunity to ride the wave of popular discontent and turn dissatisfaction towards the pro-development faction in their favor. A big blow for the government came after 18 scholars and academics belonging to the prestigious research institution *Academia Sinica* (中央研究院 *zhongyang yanjiuyuan*) signed a petition against the planning of the petrochemical complex and its construction. Given the outstanding international reputation enjoyed by the institution and the issues raised by these 18 scientists, the government could not simply ignore their complaints (Chou Chang-hung, interview, June 30, 2011). Their main concerns dealt with the destruction of the wetland, the ensuing risks for residents and the future unlikelihood of meeting the demands of the Kyoto Protocol to reduce carbon emissions, of crucial importance for a consensus-hungry country such as Taiwan.

The grouping of different social classes, representing a varied range of constituencies, increased the efficacy of the demands of the anti-Kuokuang faction – coupled with the pooling of a broad range of expertise and scientific knowledge that was not easy to dismiss – and made activists' arguments much more convincing. Apart from employing the available legal channels, activists resorted to a variety of creative techniques. The land trust campaign for instance, even though it did not go through due to the government's refusal to grant social activists permission to buy the land, nevertheless reached its goal by bringing several people to join in the opposition, adding more pressure to the government in a campaign year.

The KPT case illustrates how an effective framing of environmental threats can motivate a wide range of different actors to engage in a protest. Specifically, the participation of more experienced stakeholders from urban areas provided villagers with information about the impending environmental and cultural threat to their locality. The environmental hazard was made more vivid by the depiction of another environmentally damaging project, the sixth naphtha cracker, whose "side effects" on the land and the people had been known for years. The powerful rhetoric of the destruction of Taiwan's native traditions, such as its oyster fisheries, constituting a fundamental element of the island's cultural origins as a maritime land and tied to several folk religion elements, such as the

Mazu cult, created a sense of outrage and injustice among villagers that triggered a heartfelt and widespread participation.

The piggybacking on other issues was used to take advantage of the specific political opportunities available at that given moment. In fact, in 2010 the central government was beginning to eye with discomfort all sources of popular unrest, for fear that these may damage its reputation and ruin its chances in the upcoming presidential election of 2012. Activists' choice of involving a team of legal experts in the case, who made linkages to various contentious issues such as land-related claims, had the immediate effect of drawing island-wide visibility to the case, which was exactly what the central governmental was trying to avoid.

Various factors contributed to the formation of this piggybacking strategy in Taiwan's environmental struggle. First, environmental laws are quite advanced but often do not succeed in effectively tackling problems due to a still overarching influence of the pro-business faction. It is often difficult – if not impossible – for victims of pollution and resource-poor farmers and fishermen to obtain valid evidence of industrial pollution and its consequences on both human health and the ecosystem relying solely on themselves. Second, many polluting factories are built on land that has been obtained with under-the-table deals. Therefore, protesters have a more solid legal argument if they base their claims on land-related disputes rather than trying to establish a causal link between pollution and health and/or ecological impacts. Third, land regulations are quite straightforward and it is easier to prove any violation and/or abuse. For this reason the legal team involved in the KPT case based its claims almost exclusively on land issues, by identifying those clauses in the various land-related regulations that had been violated by both the developers and local and central authorities' conduct.

Owing to the mounting island-wide pressure towards illegal land-deals and requisitions (for instance in the above-mentioned well-known case of Dapu, in Miaoli), the central government feels required to enforce land regulations (Hua, 2010: 60–61), especially in politically crucial moments such as in a pre-election year. Therefore the upcoming political elections created several fruitful opportunities for environmentalists all throughout 2010–2011, due to the heightened awareness of environmental issues after Fukushima; hence political parties were forced to second mounting popular discontent against the facility, by jumping on the environmental bandwagon, to gain wider popular support.

It is very plausible that if the climax of the protest towards KPT had taken place in a less sensitive year, its fate might have turned out to be very different. Thus green activists' ability also resided in their capacity to fully exploit the strategic moment of the electoral campaign, during which they turned KPT into an issue of crucial importance with both candidates, Ma Ying-jeou and Tsai Ying-wen, compelled to express their opinions on the project.

Conclusions

The KPT success consisted, among other things, in the skillful mobilization of Changhua villagers' "local identity" perception to frame industrial investment as a threat to their traditional way of life and to the surrounding natural environment. Local identity thus became a powerful weapon, strengthened by the strategic moment in which the protest took place, with politicians campaigning for the upcoming presidential election. Fishermen and farmers were concerned with their livelihood prospects and future income and dissatisfied with the compensation plan put forth by the authorities. The KPT case embodies the perfect example of environmental activists having learned how to mix scientific and lay knowledge in their action repertoire to fight the state. As I have shown, with the help of several concerned NGOs and a few legal experts, they were able to gain admission to roundtable discussions with experts and EPA officials. In return they obtained access to privileged information in regards to the possible long-term effects on the environment and on public health. This attitude of challenging state expertise and state-set procedures for environmental assessments, through those very same institutional channels established by the state, shows that the employment of rules-based participatory approaches is becoming the norm. While in the past villagers often discarded environmental concerns as soon as satisfactorily monetary compensation materialized, this case symbolizes the beginning of a new era for Taiwanese modern environmentalism, with the protest comprising people and associations from all over Taiwan rather than merely involving locals and their parochial interests. Residents of the concerned townships, rather than assuming an attitude of distrust towards outsiders, as was the case in many protests of the past, gladly accepted outside help; this connection gave a much broader appeal to their claims, making their tactics more globalized. Lastly, protesters and activists secured this "victory" entirely on their own, without political parties' help; the DPP was in fact only a latecomer, trying to take advantage of the rising tide of discontent against the project, rather than a true supporter of the anti-Kuokuang movement from its onset.

Notes

1 According to the official statistics of the Fishing Industry Bureau (漁業署 *yuyeshu*) in 2008 the quantity of oysters farmed reached 3,030 tons for a total financial output of NT$230 million (Lin, L.H., 2010: 20).
2 Naphtha normally refers to a number of different flammable liquid mixtures of hydrocarbons, i.e., a component of natural gas condensate or a distillation product from petroleum.
3 KPT Co is actually an affiliate of state-run oil refiner China Petroleum Corporation Taiwan (台灣中油 *Taiwan zhongyou*) and the project in Changhua was aimed at expanding the country's oil-refining capacity and the production of chemical products such as ethylene.
4 Taiwan's second biggest wetland, namely the wetland of Dacheng (大成溼地 *Dacheng shidi*).
5 Currently fewer than a hundred Taiwanese white dolphins remain; their population has plummeted in the past few years due to an increase of polluting factories nearby

the coastal areas. They now face extinction caused by pollution, habitat encroachment and widespread public indifference (Sui, 2011a). Due to these reasons, the white dolphin, characterized by the pinkish hue of its skin, has become the symbol of the protest towards the petrochemical industry in the area and countless posters and blogs have been dedicated to it (see, for example: http://taiwansousa.blogspot.com/, last accessed on February 9, 2015). (See Figure 4.2).

6 The petrochemical industry refers to the manufacture of all chemical products involving natural gas or petroleum, ranging from upstream basic petrochemical feedstocks; midstream petrochemical raw materials as well as downstream petrochemical products.

7 Two weeks after the lifting of martial law, villagers in Houchin started a fierce protest against CPC's plan to expand a refinery installing the fifth naphtha cracker. The protest lasted three years, and was then terminated due to a variety of reasons ranging from material resources depletion to inside factionalism among protesters. Thus, in 1990 the delayed construction could resume and in 1994 the fifth naphtha cracker finally became operational (Ho, 2005c: 239).

8 Furthermore, as noted by Ho (2010b: 452), previous similar "incidents," such as the case of the Linyuan Petrochemical Industrial Zone, where less water led to less agricultural production and wastewaters, harming the local aquaculture business, now serve as an example of what could happen, for locals in Changhua. Since immediate offshore areas would have been contaminated by chemical materials fishermen would have had to sail further for their catch to be less polluted, while their sales would have risked, as was the case in Linyuan, being rejected by wholesalers for fear of pollution.

9 In order to boost water and soil conservation in Changhua and Yunlin counties, among Taiwan's areas the most affected by land subsidence, the Council of Agriculture in 2013 set up a plan, titled *Golden Agricultural Corridor Plan*, which establishes afforestation and dry farming areas spanning 1.5 km on both sides of the high-speed rail line. Nevertheless, the Changhua County government is currently planning to expropriate more of such plots of land that shall be used for the establishment of a cluster of firms dealing tires (Lin, 2014).

10 This part of Taiwan constitutes an industrial belt of more than 100 km in length, which stretches from the Dajia to the Jhuoshuei river making the wetland between Dacheng and Fangyuan the only unspoiled piece of land left in the area. Further concerns regard the EIA approved by EPA in 2009, which allowed the science park to discharge its industrial wastewater into either the new Jhuoshuei River or the old Jhuoshuei River (Lee, 2013a), thus potentially threatening local farmers' livelihood by polluting the scarce water they have left.

11 Experts and scientists speculated that DEHP consumption from an early age might be particularly harmful, leading to a feminization of males or to masculinization effects on females (*The Economist*, 2011).

12 Unfortunately Taiwan's history with food-safety scandals has been dotted by a crescendo of negative stories in the past few years, which reached its highest peak in September of 2014 with the sadly famous gutter-oil scandal that was portrayed by several international media outlets (*New York Times*, 2014).

13 The report concluded by highlighting the health hazards provoked by the petrochemical industry (Jhan, 2011: 48–54).

14 After the Kyoto Protocol became effective in 2005, even though Taiwan is not a formal signatory to the pact, it has aimed at increasing the amount of electricity needs derived from renewable sources, among other reasons, in an attempt to boost the country's international status and recognition in view of its thorny position vis-à-vis Mainland China (Shih, 2005) and its undefined status. The overall goal is to cut its carbon dioxide emissions by a total of 170 million tons by 2025.

15 Ethylene is a lucrative and much-sought-after petrochemical product used in construction materials, textiles, plastic and film products. Critics of the KPT project claim it

does not make sense to emphasize Taiwan's self-sufficiency in ethylene since to create it, you first need to import crude oil and refine it through a costly process (Wu Kuen-yuh, interview, September 20, 2011).

16 For an integral viewing of the video see: www.civilmedia.tw/archives/4904 (last accessed on February 9, 2015).

17 Taiwan Environmental Protection Union (TEPU) has ten local chapters of which one is located in Changhua County. For more details see: http://tepu.yam.org.tw/intro.html (last accessed on February 9, 2015).

18 The KMT approved the appointment of vice president (since 2012) Wu Den-yih as the party's first vice chairman in 2014, making him the ROC's second most powerful man after President Ma. From 2009 to 2012 he served as premier.

19 Available at: 白海豚之歌 www.youtube.com/watch?v=6YzOIqUlISI (last accessed on February 9, 2015).

20 He was so adamant that I would fully understand the great extent to which politics played a role in this case that he invited me to stay at his "tree house" for one night, so that his daughter and himself could explain the complicated behind-the-scenes in regards to KPT.

21 According to Article 14 of *the Environmental Impact Assessment Act*, all projects of a certain size must first go through an Environmental Impact Assessment.

22 Land-use regulations in Taiwan are purposely kept vague so as to provide officials with enough maneuvering space when evaluating developmental projects, and keeping public involvement at the minimum. For more on this see Tang and Tang (1999: 358).

23 Preparatory document of the plaintiffs n. 26. May 2009. On file with the author (courtesy of lawyer Lu Shi-wei). A lengthy dispute surrounds the existing policy of shipping sand and gravel from eastern Taiwan, for construction purposes, to the more industrially rich western plains.

24 According to an estimate of the Directorate-General of Budget, Department of Accounting and Statistics, the whole petrochemical industry of Taiwan contributes less than 4 percent of the country's GDP and if one were to analyze data from previous similar facilities such as the sixth naphtha cracker in Mailiao, one would realize that in 2008 and 2009, 60 percent of gasoline, 90 percent of diesel fuel and 5 percent of ethane were exported (Official Cost-Benefits Analysis of KPT, courtesy of Prof. Chen Chi-chung). That means that the products of petrochemical industries are mainly meant for export but not for domestic demand thus fueling activists' and opponents' opinions that Taiwan's soil and citizens' health are being sacrificed for other countries' gains.

25 A few months after the final decision to cancel KPT, the idea of building the petrochemical complex in Changbin's Coastal Industrial Park (彰濱工業區 *zhangbin gongye qu*) circulated with a certain insistence. Changbin Industrial Park, in Changhua, is Taiwan's largest, with a total size of 3,643 hectares. However, this idea has been shelved in the meantime; as of July 2012, Taiwan's Kuokuang Petrochemical Technology was expected to revive its plan to build the integrated refinery in Pengerang, Malaysia (*Taipei Times*, 2011) but in 2013, as CPC had completed its feasibility study, it decided not to proceed with the project.

5 The Taipei Dome BOT Project

Background information

The "Taipei Big Egg" (臺北大巨蛋 *taibei da judan*), as the Taipei Dome is known in the city (to mark its difference with the already existing "small egg," aka the Taipei Arena), is a long-standing project whose planning (at the current location) dates back at least to the year 2001. The Taipei Dome complex, scheduled for completion in 2015, will occupy ten hectares of land and will include a 40,000-seat indoor stadium, hotels, department stores, a shopping center and office buildings. The Dome is being built at the site of the abandoned Songshan Tobacco Factory in Xinyi District,[1] a beautiful legacy of Japanese colonial history dating back to the 1930s.

Figure 5.1 Taipei Dome at the early phases of its construction, August 5, 2011 (source: Simona Grano).

Figure 5.2 The Songshan Tobacco Factory, October 2, 2011, Taipei (source: Simona Grano).

Figure 5.3 The Songshan Tobacco Factory, October 2, 2011, Taipei (source: Simona Grano).

To be precise, the history of this developmental complex harks back to the 1990s, when the birth of a professional baseball league increased the desire to have an indoor stadium (Chen, 2009b: 86). Many years passed without a suitable site being found. Eventually, as Ma Ying-Jeou served his term as Taipei Mayor (1998–2006) the final decision to build the large dome on the site of the Songshan Tobacco Factory reached a consensus. In 1998, due to a variety of factors such as the end of the state monopoly on the cigarette and alcohol markets, the factory had closed down and its pricey park, 18 hectares located in glamorous East Taipei (nearby the famous 101) fell into a state of decay (Chen, 2009a: 74). Several plans were drafted, while the city government pondered what to make of this vast area; the prevalent idea was to turn the Songshan factory into a cultural park for young generations, while preserving the splendor of the historical site. For these reasons, after the decision to build the dome on the factory grounds was reached, the Council for Cultural Affairs decided to preserve 8 hectares, including the actual Tobacco Factory building, and turn them into a cultural park. The residual ten hectares would be used to build the Taipei Athletic Park Plan (or Taipei Dome). Therefore, the 18-hectare Songshan Tobacco Factory has thus been divided into a cultural park and a sport park, whose bid has been awarded to Farglory Corporation (園雄集團 *yuanxiong jituan*) (Our Island, 2014).

The first Environmental Impact Assessment process for the project was concluded in 2003 and gave the project a green light. It was decided that the Dome would fall under the BOT model.[2]

Figure 5.4 Taipei Dome's neighborhood, October 2, 2011 (source: Simona Grano).

At the initial planning phase everything seemed to proceed smoothly, from the re-zoning of the land to the relocation of local residents; then in 2006, after the city government signed a 50-year BOT contract with Farglory and began transplanting 800 trees (Chen, 2009b: 86) residents and citizens finally began paying attention to the Sport Dome project, which sparked numerous, heated, discussions. While social organizations worried mostly about the environmental impact to the factory's park and the damage such a project could cause to a cultural heritage vestige like the Tobacco Factory, local residents were mostly concerned about the increase in traffic and the higher levels of noise and smog, which such a complex would certainly generate (Our Island, 2014). The big number of trees sacrificed for the project brought several activists onto the streets while numerous "tree-hugging"[3] campaigns have been organized in Taipei since 2011; activists' distrust towards local authorities – deemed guilty of merely paying lip-service to environmental protection for electoral purposes – has been amplified by the numerous instances in which trees were felled to accommodate the urban development plans of the city government. At the end of 2006 680 trees, located in the Songshan Tobacco Factory grounds, were transplanted to the future site of Nangang's Baohu Junior High School, waiting to be replanted elsewhere, as soon as a suitable site could be found. More than 100 trees, which were outside the dig area, were left in peace. One hundred and sixty other trees were "temporarily stored" in the cultural park lot of the Songshan Tobacco Factory site, with the idea of returning them to their original place, in the sporting area, once the Dome is completed. In 2007 local citizens discovered that one-third of the trees that had been transplanted to Nangang Baohu Junior High had already died (officials of the Nangang park list the number of dead trees at 103). Hence, citizens and environmental groups decided to hire some experts to determine what had happened and whether any of the dead trees were actually older than 50 years and should therefore have been protected by the existing trees' protection regulations. Before their study could begin, they were told that the park's officers had already removed all the dead trees with the explanation that they were trying to avoid the spread of diseases to the rest of the healthy trees. Five other old giant trees, including a camphor tree, were transplanted to the cultural park with the idea of leaving them there for fear that repeated moves would prove to be fatal. The camphor tree died shortly after but its trunk was kept as symbolical homage to the tree (Pan, 2011a: 8). In 2009, on 228 Memorial Day,[4] six environmental activists and Green Party members decided to form a circle around the trunk of the old camphor tree. Shortly after they were arrested and detained for a few hours although, in the end, prosecutors decided not to press charges citing the protesters' rights to freedom of expression. During one of our interviews Pan Han-sheng, founder of the Tree Party, shared some information on the growth of a Chipko movement[5] in Taiwan and how, recently, numerous attempts to fell trees on the part of local and central authorities are making this movement increasingly popular among activists with ensuing growing interest from the media.[6] Pan revealed how in the previous years the city government had transplanted more than 300 trees, placing them in "tree banks" (Pan Han-sheng, interview, September 1, 2011). These places, designed to be a new home for displaced trees, turned out to be their graveyards as most of the hastily transplanted trees did not survive (Our Island, 2011b).

Irregularities and problems regarding the Taipei Dome

At the beginning of 2008, due to some changes in its plan, Farglory submitted documents for a second planning review and for a second EIA. However, construction was nonetheless started before this second EIA could reach its completion.

Several other problems that have marred this project along the road have remained open controversies until today.

First of all, activists complain about the authorities' non-transparent behavior in approving the Big Dome project. The Committee responsible for the EIA approved the project on May 26, 2011 but it named a few conditions that should have been satisfied before the project could take place (Our Island, 2014). These conditions are: the total surface area dedicated to the so-called "auxiliary activities" under the Sport Dome BOT (namely: hotels, office buildings, cinema complex and so on)[7] should have been reduced by 17.4 percent to 202.610 m^2 while parking spaces should have been increased to 187.965 m^2 (Mo, 2011a); further, Zhongxiao East Road will have to be expanded from a four-lane street into a five-lane road; the separation isles will have to be removed and some of the pedestrian sidewalks will have to be turned into car lanes; these last provisions have been carried out, starting from the summer of 2014.

Afterwards the case was sent to the Taipei Urban Planning Commission, which, after a short period of consideration approved the project on June 16, 2011, provided that Farglory would meet those provisional conditions before starting the actual construction phase (*The China Post*, 2011c). Needless to say, activists and opponents lament that the conditions were not satisfied before actual construction began. Furthermore, aware that in order to enlarge Zhongxiao East Road many more trees would have to be felled, borough warden of Xinrenli 新仁里 (in Xinyi District), Li Tsai-Jiu (李財久) launched a petition in 2011 to ask the government for a detailed account of the process and expressed his disapproval in regards to the widening of the road, pleading for the trees to be spared. After more than three years, as of 2014, borough warden Li is still waiting for an answer; Farglory's Dome, on the other hand, has already started to take shape.

According to a transportation expert: "A large stadium that can accommodate more than 10,000 people needs to be carefully planned and possibly set in an area where traffic flow is not severe and where evacuation can be dealt with swiftly" (Chang Sheng-hsiung (張勝雄), interview, September 24, 2011). Even though the Taipei Dome will be located in the immediate vicinity of a metro station (the Sun Yat-Sen Memorial Hall station), which can accommodate a maximum of 4,700 people at once, it has been calculated that, with a train arriving every four minutes, it would take more than one hour to dissipate the crowd in case the stadium is filled at its maximum capacity (Peopo Curiosity Baby,

Figure 5.5 Taipei Dome in early 2014 (source: Rosa Enn).

2011). Furthermore, due to the location of the Dome in Taipei's bustling financial district, the rising flow of buses and private cars would certainly contribute to worsen the already problematic traffic situation while increasing smog levels.

Further, critics of the project maintain that there is no need for such a stadium; since sporting events of a certain caliber are quite rare in Taiwan, the Dome would mostly be used for musical performances, for which there are already more than enough locations (Pan Han-sheng, interview, September 1, 2011; Arthur Yo, interview, October 21, 2011). The phenomenon of developers building increasingly costly venues, residential buildings and other facilities, which then lay empty and derelict is quite common in Taiwan. Since this is a BOT project though, citizens fear that for the government it will be nearly impossible to regain its costs thereby damaging taxpayers who, ultimately, are the sole contributors to this endeavor (Arthur Yo, interview, October 30, 2011).

Furthermore activists lament the low percentage of pro capita green space in Taipei – 5.16 m^2 vis-à-vis Beijing's 12.6 m^2 or Seoul's 15.93 m^2 (Lee, 2011b) – and urban heat, which is on the rise with more and more days per year displaying a temperature over 35°; a green space would help to lower urban heat and modulate rainfall (Robin Winkler, interview, September 5, 2011).

Until mid-November 2011, green activists had hoped that Mega Bank Holdings would refuse to grant the money for the investment since the returns will almost certainly be low. The fact that the bank did grant the loan despite the grim financial prospects has convinced activists that there has been governmental influence in pressuring the bank to lend Farglory Corporation the sum for the project.[8] "Proof of this", they claim, "can be found in the fact that 25% of Mega Holdings' shares are in the name of government officials while more than 70% of directors come from the Ministry of Finance" (Green Party Press Release, 2011). To sum up, green activists and opponents are keen on highlighting the fact that: "The Dome BOT is a profit driven project, whose main scope is building commercial facilities masked as sporting facilities" (Pan Han-sheng, interview, September 1, 2011). In comparison to the KPT case, whose extensive impact and activists' "noble goals" (saving endangered species and the ecosystem) attracted widespread media visibility and heartfelt participation and support from all over Taiwan, in the case of the Taipei Dome the impact is localized and considerably fewer people got involved. Therefore, as we shall see in the next section, piggybacking on other issues to take advantage of specific political opportunities was the only available strategy.

A further explanation, accounting for the lack of heartfelt participation in the anti-Dome campaign and the resistance encountered by the green movement in soliciting support for its anti-Dome campaign, is directly related to what Tang Chin-Ping views as an urban political system that has "features of both the pro-growth urban regimes of Western democracies and the clientele network of an authoritarian regime" (Tang, 2003: 1029). In fact, urban environments have partly turned into elite-governed areas, where socially disadvantaged or non-politically influent groups' interests (such as social activists) are often under-represented. In fact the main supporters of the Dome Project are the Taipei City government and the central government; the latter due to Ma Ying-jeou's former role as Taipei Mayor with the Dome project having been decided during his tenure. Activists consider the government's stance "non-transparent" at best, due to the close relationship between vice-president Wu Den-yih and Chao Teng-hsiung, chairman of the Farglory Corporation.

Actors involved and methods of contention

Like other Taipei residents, a group of writers, lawyers and architects came to know the Songshan Tobacco Factory park after the lifting of the state monopoly on the tobacco industry towards the end of the 1990s, as the factory closed down and its pricey park in downtown Taipei was left untouched. Interested in the potential of such a green space and the beautiful colonial building of the factory, they were convinced of the need of turning the place into a cultural park. As mentioned before, their wishes were short-lived since eventually, as Ma Ying-jeou served his term as Taipei Mayor (1998–2006) the final decision to build the large dome on the site of the Songshan Tobacco Factory reached a consensus. After holding a few public protests of small scale in 2011 they established the *Songshan Tobacco Factory Park Union* (松菸公園催生聯盟 *songshan gongyuan cuisheng lianmeng*), with Arthur Yo – You Yi – (游藝) as its chairman. A strategic alliance coalesced, made up of several environmental groups, advocating for ecological sustainability and turning the area into an ecological/cultural park. Together with other concerned activists, with the Green Party of Taiwan and one legal association, this assembly came to comprise several anti-Dome groups: *Organization of Urban Re-s – OURs* – (專業者都市改革組織

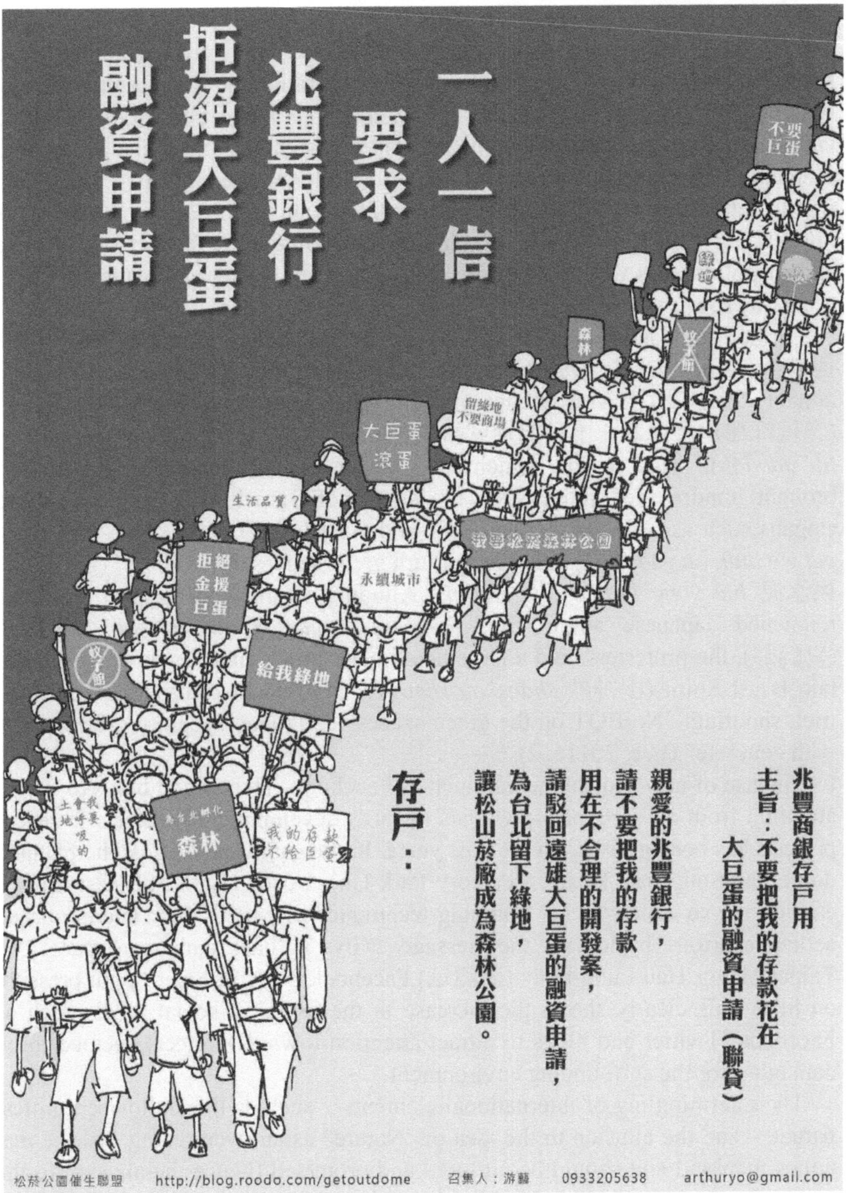

Figure 5.6 "One man one word" petition poster.

zhuanyezhe dushi gaige zuzhi), the *Taiwan Environment Information Center* (環境資訊中心 *huangjing zixun zhongxin*), the *Society of Wilderness* (荒野保護協會 *huangye baohu xiehui*) and the *Wild at Heart Legal Defense Association* (蠻野心足生態協會 *manye xinzu shengtai xiehui*); together they held several activities in 2011 aimed at emphasizing the importance of living in harmony with nature, saving a cultural legacy such as the Tobacco Factory and increasing Taipei's share of green areas.

First, they organized a petition called "One man one word requesting Mega Bank Holdings to refuse granting Farglory Corporation the loan for the Big Egg BOT" (一人一言要求兆豐銀行拒絕大巨蛋融資 *yi ren yi yan yaoqiu zao feng yinhang jujue da ju dan rongzi*); the petition was signed by more than a hundred academics, specialists, lawyers and numerous literati (Lee, 2011b). Their goal was to render the BOT's negative effects known among the general public: its impact on a historical building and on the surrounding trees, the increase in noise and air pollution as well as the traffic jams that would result in downtown Taipei (Pan, 2011b).

Furthermore, activists organized a Halloween Parade Protest (萬聖節的抗爭運動, *wanshengjie de kangzheng huodong*) called "Protect Green Spaces, Sustainable Taipei – If you do not give us green spaces we break everything" (守護綠地永續台北－不給綠地就搗蛋 *shouhu lüdi yongxu taibei – bu gei lüdi jiu daodan*).[9] This parade, which saw the active participation of the author, brought hundreds of activists on the streets of Taipei with people shouting slogans such as "We want green areas and not BOTs" (給我綠地，不要 BOT *gei wo lüdi, bu yao BOT*) and "Do not turn green spaces into cement" (別用綠地換水泥 *bie yong lüdi huang shuini*). Alluding to the Forest Spirit from the renowned Japanese anime film *Princess Mononoke* by Hayao Miyazaki (宮崎駿), the protesters held a large hand-made green puppet named the Mountain Beast Spirit (山神獸 *shanshou shen*) as they marched through Xinyi District, shouting: "No BOT on the green spaces" and "Don't replace green spaces with concrete" (Lee, 2011c: 2).[10]

The use of new "social media" such as Facebook, Twitter and blogs to attract attention from all sorts of individuals towards developmental projects in recent protests has been growing in the past years. In the campaign against the planned dome, the Songshan Tobacco Factory Park Union called on Taipei residents to employ the so-called "cyber bombing technique" as part of their internet-based action repertoire by leaving the message: "Give us back our green space," on Taipei Mayor Hau Lung-bin's (郝龍斌) Facebook page, in order to put pressure on him. This clearly shows the increase in the usage of social media such as Facebook, Twitter and blogs to attract attention towards projects deemed to be damaging for the surrounding environment.

The intermingling of international elements – such as the Halloween protest parade – and the allusion to the idea of "Nature" as an overarching essence that unites mankind and should be nurtured and protected (hence the allusion to the Forest Spirit) connects the globalized cultural dimension of Taiwan's environmentalism with purely traditional and cultural elements, such as the protection of

Figure 5.7 Halloween parade, October 30, 2011, Taipei (source: Alexandra Baraille).

the island's peculiar colonial heritage (embodied by the Songshan Tobacco Factory grounds).

Lastly, green activists organized numerous protests and press conferences in the hope of attracting media attention; a particularly successful one involved a theatrical representation of the tight and non-transparent relationship between

Figure 5.8 Halloween parade, October 30, 2011, Taipei (source: Simona Grano).

Figure 5.9 Pan Han-sheng, founder of the Tree Party, first from left, at the Halloween parade, October 30, 2011, Taipei (source: Simona Grano).

Figure 5.10 The puppet of the Mountain Beast Spirit, October 30, 2011, Taipei (source: Simona Grano).

then Premier Wu Den-yih and Farglory's chairman (Green Party Press Release, 2011) as well as using creative ways to enrich their collective action repertoire such as "planking activities" as a form of protest (Zhong, 2011).

Critics of the project believe that the Dome will harm the neighborhood in several ways. Nevertheless they are aware that it will be hard to make a strong

Figure 5.11 "Planking" as a form of protest for the protection of green spaces, October 2, 2011, Taipei (source: Simona Grano).

claim for environmental damages, due to the limited impact of the Dome, in comparison to a project with a far wider extent such as a petrochemical plant or nuclear facility, for instance. Based on this understanding, all throughout their most active protest phase, they tried to piggyback on other issues, such as the protection of an historical building, like the Tobacco Factory. This attempt failed on the grounds that the factory had in fact been preserved and only its vast park size had been cut down. When I interviewed Professor Lin Tze-luen, a professor of political science at National Taiwan University, with a past of student activism, he shared his opinion in regards to a possible exit from this situation of stalemate:

> Maybe if one can show the developers that you can make money and earn a profit also by "greening a city," this strategy could successfully be employed for the Dome in order to turn the old tobacco factory into a green park, rather than a controversial music arena with annexed shopping mall.
>
> (Lin Tze-luen, interview, July 15, 2011)

The political element

In the anti-Dome campaign, the strategic temporal dimension of the protest taking place in a pre-electoral year was again highly significant; equally important was the indirect role played by the Green Party. In fact, 2011 was a pre-electoral campaigning year for both the presidential as well as legislators' elections of 2012; in the latter, the Green Party of Taiwan and the DPP cooperated for the first time, precisely in the Xinyi-Songshan District. Here, the GP nominated Pan Han-sheng (who meanwhile left the GP for the Tree Party in August 2014) as its candidate so that the Pan-Green Alliance, instead of nominating its own candidate, shifted its weight compact behind Pan. The official explanation for this cooperation, offered by the DPP, was that the two parties would cooperate on a "policy basis," meaning that they had several common targets, such as reversing the country's nuclear power strategy and, on a smaller scale but specifically related to Xinyi District, stopping the construction of the Taipei Dome. However, the cooperation was not smooth and several problems arose, directly related to the different supporters' base of the two parties.[11] For the Green Party the Taipei Dome case proved to be one of the crucial cornerstones of its electoral campaign since it embodied all the relevant characteristics and issues that are essential for hard-core Green Party supporters around the

Figure 5.12 Green Party weekly study session, October 20, 2011, Taipei (source: Simona Grano).

world. The GP's tactics during the pre-electoral period were not so much focused on personal contacts and social capital (shaking hands, attending weddings and funerals – as is common in the repertoire of better established and more traditional parties in Taiwan) but more on spreading environmental awareness; therefore the importance of the Taipei Dome case, which functioned as the perfect model to promote the GP's ideology, is not to be underestimated. The GP's activities, as is the norm for all Green Parties around the world, tend to be of high cultural value and range from weekly study groups (I often took part in the study sessions in Taipei), during which a plethora of contemporary topics are discussed, to posting articles on the internet, organizing awareness campaigns (offline and online) and handing out leaflets.

Currently, the Green Party of Taiwan is hardly influential and in great need of an image boost; this was one of the decisive factors for opting to cooperate with the much more powerful and well-established DPP in 2011. The GP, itself not particularly skilled in canvassing techniques, profited out of the close collaboration with the DPP, which in the Xinyi-Songshan District was campaigning for Tsai Ying-wen as president and Pan Han-sheng for legislator. The association between the two was fairly successful and quite a few DPP supporters mistook Pan for one of the DPP's own candidates (Pan Han-sheng, interview, September 1, 2011). On the other hand, for the more radical and less ready to compromise GP's electorate – usually positioned along a deep-ecology scale of environmental ideology[12] Pan's closeness with the DPP was not always welcomed as it came across as "too compromising" and thus as a "selling out of environmental interests." Conscious of this, Pan Han-sheng had to adjust his strategy during the campaign several times and became, in the end, quite selective in his choice of which DPP rallies to attend and which not; towards his own electorate he tried to justify the collaboration by emphasizing the common goals of the two parties and the benefits the GP could derive from this selective policy-cooperation (Pan Han-sheng, informal communication, November 10, 2011). In turn, this "distancing-strategy" angered several DPP supporters, who thought of him as "exploiting the DPP," without giving anything in return. The basis for the cooperation between the two parties rested on environmental conservation – or at least this is the side that the GP was keen to emphasize – while for the DPP the main issue for the teamwork was the common anti-nuclear positioning. The GP focused its media campaign on the need for policies that would curb nuclear power and on the loss of green spaces in urban Taipei; in this regard, the destruction of the Songshan Tobacco Factory's grounds became the symbol of the GP's core values, when running for election in the Xinyi-Songshan District. Pan's canvassing tactics were in line with his political ideology and both while protesting against the construction of the Dome as well as campaigning for his election, he refused to use motorized vehicles and reduced the number of electoral banners to minimize the ecological impact of his campaign (Lai Fen-lan, informal communication, June 30, 2011).

In view of their common anti-nuclear stance, both presidential candidate Tsai Ying-wen and legislative candidate Pan Han-sheng were presented in turn as "the anti-nuclear president" and "the anti-nuclear legislator" to further emphasize the closeness of their respective positions. In light of all that has been said, it is easy to understand how, for the GP and for Han, the Taipei Dome case was not simply one of several protests on behalf of the environment, but rather a symbol of all that the GP strives to achieve on a national scale. This is directly linked to the issue of framing one's grievances so as to attract consensus. In fact, Pan had to shift his strategy and priorities, according to which response he seemed to be getting from its supporters, several times during the course of his campaign. Furthermore, hampering the GP's strategic campaigning efforts was the competitive relationship which existed with other movement leaders, above all, Arthur Yo, the previously mentioned leader of the main NGO involved in the protest and a local resident of Xinyi, who did not see eye to eye with the GP's strategy (Pan Han-sheng, interview, September 1, 2011; Chan Shun-kuei, interview, November 18, 2011). Lawyer Chan Shun-kuei remarked, during one of our encounters, that he had several times offered his help to Arthur Yo but was rebuked on the grounds that Yo and his organization did not agree with his strategies and ideas to fight the sporting facility (Chan Shun-kuei, interview, November 18, 2011). The vicious factionalism, devised in the opposition camp, split green activists, weakening their efforts. Thus, the different actors involved, rather than cooperating among themselves, struggled with the state and with each other, for supremacy. In such a situation, those agents occupying a more dominant position and with more means at their disposal (namely governmental officials and the developers) gain the upper hand by exploiting the other faction's internal strife, as was the case in the Taipei Dome.

Rules-based participatory approaches: the EIA process

In any major environmental battle taking place in contemporary Taiwan, protesters now have a selection of participatory approaches at their disposal, through which they can effectively take part in the decision-making process, concerning developmental projects. As shown in Chapter 1, since 1994 projects of a certain size (at least ten acres) have to undergo an Environmental Impact Assessment before obtaining a green light (Ho, 2004); just like in the eighth naphtha cracker case, common citizens and activists actively took part in the proceedings.

Throughout the various EIA stages, environmental activists sent legal experts and volunteers to various public hearings and city council meetings – so as to lobby on behalf of their cause – and to EIA hearings, to counter attack the developers' official version through knowledge-based information, provided by their own team of professional experts.

On October 30, 2011, I interviewed one of these experts, Dr. Tsai Ya-ying, a young lawyer/activist working for Wild At Heart, who steered me through a complicated maze of legal caveats that opponents have attempted to use to fight the project. A total of four lawyers (including Tsai) are formally involved in the case.

In her interview, she summarized the three main problems her team revised, regarding the construction of the Dome:

> We have moved for a repeal on three legal flaws committed by the authorities:
>
> 1. Environmental Impact Assessment Procedure (環境影響評估制度 *huanjing yingxiang pinggu zhidu*);
> 2. Urban Design Evaluation Deliberation (都市設計審議 *dushi sheji shenyi*) and;
> 3. License to build (建照執照 *jianzhao zhizhao*).
>
> (Lawyer Tsai Ya-ying, interview, October 30, 2011)

1 EIA procedure

First of all lawyer Tsai devised that there are four key points, pertaining to the EIA procedure, which constitute actual legal violations.

I Opponents contested the validity of the decision to allow the Dome project, reached by the EIA committee, claiming that the quorum of members that must be present to approve a project, was not reached. In fact, as explained in the first chapter the top body which makes the final decision in regards to the approval of a project, is the EIA Committee composed of 21 members; among these seven are government appointees/officers and 14 come from the academia. The Commission meets once a month and the minimum quorum of members who have to be present for legal validity is 11. On the day in which the project was approved only seven official members and seven from the academia were present. Since in this case the city government has a direct stake, environmentalists maintain that city officials should not have taken part in the review process because they have a conflict of interests. The Meiliwan BOT case (美麗灣 BOT 案 *meiliwan BOT an*), where the court ruled that the county government couldn't conduct the environmental assessment due to a conflict of interests, could serve as a precedent. At any rate activists maintain that the quorum was not met because only those members belonging to the academia should be counted as "valid." In lawyer Tsai's words: "We contend that since only the seven non-governmental members were legitimate, the quorum was not met. The city government has hijacked the review process while it should have recused itself instead" (lawyer Tsai Ya-ying, interview, October 30, 2011).

II On June 28, 2010 the construction case was rejected by the first round of the second EIA process on the grounds of excessive increased volume (擴增量值 *kuozeng liangzhi*) of the auxiliary facilities and the Dome's impact on traffic conditions (especially in light of the limited space available for parking purposes). The law states that if a case is rejected, another revised case (on the same construction site) can be submitted, but the new proposal needs to solve the previous issues. Since the second submission further reduced parking spaces (therefore not meeting the requirements of an auditorium with a capacity of 40,000 people and a shopping mall, which will be built on this site as well), activists maintain that the BOT case is forging ahead illegally. Furthermore, the second project has indeed cut back on the volume of the auxiliary facilities but these are, according to critics of the project, still excessive.

III Wild at Heart's lawyers argued that the second EIA submission required a second phase; in fact, the law dictates two rounds of environmental assessment if the initial one indicates significant impact on the nearby areas. Green activists emphasized that the historical building nearby (the Songshan Tobacco Factory) is fragile and already slightly subsided; it might not bear the ground shaking caused by the construction process. In addition, there's an elementary school close by. Construction of the Dome requires 18 months of traffic flow involving dumping trucks loaded with heavy gravel for 20 hours a day, meaning 1.1–7 min per truck on average. Therefore the Committee should have requested a second phase for the second EIA submission, taking into consideration the amount of dust particles released in the air to the detriment of kids' health.

IV The decision-making process was careless. The case was passed despite some members' objection that the formula to calculate the flow of visitors was erroneous. According to lawyer Tsai and her team, the bidders simply averaged out the volumes from the first and second submissions.

2 Urban Design Evaluation Deliberation

The second legal breach devised by activists and their lawyers concerned the Urban Design Evaluation Deliberation procedure, whose committee eventually approved the project in June 2011.

I Just like for the EIA procedure, activists believe that the city government should have recused itself from the Urban Design Evaluation Commitee, to avoid any conflict of interest. Therefore environmentalists claim that the quorum was not reached when the committee met because government appointees, having a stake in the project, should have recused themselves.

II In addition, there was an internal strife since two local borough wardens (里長 *lizhang*) had tried to have one of the committee members recused. Their motivation for wanting to exclude that member was that he was also a standing member on the board of the MRT company which, just like the city government, also has a stake in the Dome project. According to Article 33 of the *Administrative Procedure Act*, if a member is asked to recuse him/herself, the process must be halted before any decision is reached. In lawyer Tsai's words: "In this case they just rolled over us" (Tsai Ya-ying, October 30, 2011).

III Last, the usage of the planned facilities was in conflict with the previously stated purpose. In fact, the project was originally presented as a sporting/cultural park with auxiliary commercial-usage buildings. Article 38 of the *Urban Planning Act* specifies that land relegated for a specific project cannot be utilized against its designated purposes (特定專用土地不得違反特定用途 *teding zhuanyong tudi bu de weifan teding yongtu*). Since in the Sport Dome BOT the commercial area is 202,610 square meters while the sport area is 122,000, this represents a conflict with the original stated purpose, which was to create a suitable venue for big sport events with barely 7.1 percent of the land being destined for commercial use (Chan Shun-kuei, interview, November 18, 2011).

3 License to build

Activists and their legal experts are of the opinion that since the EIA regarding the project is flawed, the Urban Design Evaluation, which is based on it, is consequently also flawed. The issuing of the license to build – being contingent on the Urban Design Evaluation – is therefore inconsistent.

Issues raised at one of the public hearings revealed that the Taipei city government had placed little emphasis on the Environmental Impact Assessment for this BOT project. For example, legal experts fighting for the acquittal of the Dome showed that the coordinates for areas in the Songshan Tobacco Factory grounds, destined for the protected trees cited in the EIA report were erroneous, some even indicating trees located on what was actually land that had already been ravaged by bulldozers.

Social activists and their lawyers also raised a few additional objections, pertaining to other procedural flaws. First of all, at the initial phase of the Dome's bidding contest the government released what activists consider to be misleading information, declaring the total floor area (總樓地板面積 *zong loudiban mianji*) that could be developed to be 317,356.8 square meters. Due to what developers perceived to be a fairly small plot, only Farglory submitted a bid. The volume was later raised but by then it was too late for other companies to present a bid proposal; consequently the city government was accused of insider dealing. According to the law, if the increase in volume amounts to more than 10 percent the bidding context should be declared invalid pending further investigations (lawyer Tsai confirmed that no such invalidation had taken place). Furthermore, usually the architect who maps the blueprint for a project cannot participate in the commercial development of said project, neither directly nor indirectly. In the Taipei Dome case, however, the architect kept working for a company collaborating with the developer Farglory Corporation in carrying out the construction.

Lastly, according to the *Cultural Heritage Preservation Act*, a developmental project must be halted if it impacts on a heritage site (in this case the Songshan Tobacco Factory). Social activists have chosen not to pursue this option for fear of being dismissed with the explanation that the site occupies a large area, a part of which has already been designated for preservation.

Negative outcome and later developments

As previously mentioned, the project was approved by the Taipei city government on May 26, 2011, on the condition that the new plan would cut the space dedicated to the commercial activities by 42,900 square meters – equivalent to the size of a Sogo department store – to preserve the facility's athletic mission. Songshan Tobacco Factory Park Union's director, Arthur Yo, maintains that the city government ignored the Taipei city council's decision to terminate the contract with Farglory and keep the site as a park in case Farglory would not make the improvements recommended by the Urban Design Evaluation Committee before the deadline. At any rate, construction started in October 2011 and is scheduled to finish in 2015. To effectively manage the numerous obstacles encountered by the Dome project (doubts about its usefulness and its impact on the traffic situation, to name but a few) Taipei city's Department of Education set up an ad hoc supervisory agency, the "Construction Planning Office for the Taipei Dome Complex" charged with overseeing the complex and offering the developer assistance in revising the EIA throughout its various phases (Chen, 2009b: 86); while in office this agency refused all external requests for an interview, thus hindering the chance that a pluralized and varied discussion on the part of citizens might ameliorate the project.

On April 24, 2014, with the Dome already 70 percent constructed further protests took place after more than 200 sandalwood and cotton trees – at the crossing between Guangfu South Road (光復南路) and Zhong Xiao East Road 忠孝東路 – were felled in order to make space for the widening of the avenue, thereby allowing the traffic to dispel more easily (*Apple Daily*, 2014).[13] Tree-huggers and eco-conscious activists climbed up on the remaining trees, in a last desperate attempt to protect them and sought to win more time while temporarily delaying the procedure. Local residents report that the whole removal of the trees to widen Zhongxiao Road was hurried and secretive. In fact, while the announcement regarding the removal of the trees was posted on April 18, 2014, barely three days later, on April 21 the transplant process had already begun (Our Island, 2014) and on April 23, 20–30 trees had already been cleared away, sparking several, heated, protests.[14]

To solve the impasse, on April 27, 2014 both sides launched a harmonization conference at which local residents called into question the transplant procedure of the trees, claiming that the treatment of the roots of the trees, which had to be made within three months from their being uprooted, was not performed. Apparently Farglory's contractor failed to carry out this task, thus greatly lowering the trees' survival rate. Farglory's answer was that the survival rate of the trees was higher than 80 percent (Our Island, 2014; Pan Han-sheng, email communication, June 29, 2014). Further conflict arose in August 2014 as the Ministry of Transportation (交通部 *jiaotong bu*) put forth a newer proposal for the widening of Guangfu South Road, which would require the transplanting of an even higher number of trees than those initially forecasted, sparking protests among tree huggers and environmental organizations (Lai, 2014c). The Ministry maintains

Figure 5.13 Trees scheduled to be felled to widen the roads in front of the Taipei Dome, April 24, 2014 (source: Ketty Chen).

Figure 5.14 Tree-huggers and environmentalists climb up on trees, scheduled to be felled, in an attempt to save them, April 24, 2014, Taipei (source: Ketty Chen).

that failing to uproot and remove the trees will pose a threat for Taipei's 2.8 million citizens by further slowing traffic; it also stated that its plan was drafted after lengthy discussions with several transportation experts and academics. On this point, Farglory representatives presented their own evaluation, stating that in the case the trees were not transplanted, those trees that have already been planted in the Taipei Dome area might suffer due to lack of sufficient space to bloom and that underground pipelines and water drains, disrupted by the roots, could also be affected (Lai, 2014c). Transportation expert Huang Tai-sheng (黃台生), who claims that he was not initially informed that the preservation of the existing trees might turn out to be an issue, maintains the need to uproot them, in order to allow car drivers to turn right (as opposed to simply left) on Guangfu Road thus dispelling traffic more easily. Other scholars, such as the already-mentioned Chang Sheng-hsiung, believe that a solution to save the trees could be to allow the traffic to simply go straight or turn left instead of right, as is the case for many roads in Taipei, where you can only turn in one direction. Such a detour, in the opinion of tree-huggers and eco-volunteers, rather than increasing polluting emissions, as claimed by the Ministry of Transportation, would decrease them because more people would opt for taking the metro. Social activists do not rule out more extreme measures, such as occupying the premises of the involved authorities, should more trees be transplanted elsewhere.

Furthermore, following the bribery allegation scandal that erupted in June 2014, involving Farglory's chairman, Chao Teng-hsiung, and the subsequent damage to Farglory Corporation's image, a group of activists has accused the company of wrongdoing and illicit acts in regards to the Taipei Dome. The reasoning is that if Farglory has cheated before, paying bribes to push forth several of its projects, it could have done so in the Taipei Dome case as well; therefore the city government should cancel the project, they claim, even though construction started already back in 2011.

Such recent events have spurred the anti-Dome campaign, which had been dormant for a while, back into action. In light of the legal violation that activists had devised back in 2011, Arthur Yo, with the help of another group of activists dedicated to protecting trees (Green Power Union) released a statement to the media in regards to the several flaws surrounding the process by which the developer had won the bid for the Dome, claiming that it been awarded to Farglory due to bribery and bid-rigging (Tsai Ya-ying, email communication, June 20, 2014; You Yi, email communication, June 29, 2014; Shih, 2014). Consequently, activists decided to file a petition against Farglory at the *Agency Against Corruption* for illegal acts during the bidding process (Lee, 2014e). This decision was announced during a public rally on June 6, 2014 organized jointly by the members of the *Songshan Tobacco Factory Park Union*, with lawyers/activists from the *Wild at Heart Legal Defense Association* and other environmental activists.

The anti-Dome movement claims that Farglory treated officials from the Taipei city government to expensive dinners and gifts (in one case activists even suspect that a cash payment was made to one of the review committee members) all throughout the evaluation, screening and bidding processes, as well as later on during the signing of the contract stage and, finally, during the Environmental Impact Assessment procedure. Such questions have clearly been raised in light of the recent bribery scandal surrounding Chao Teng-hsiung; most recently though activists have started questioning the construction schedule for the Dome project, which, according to the contract signed between Farglory and the city government, was quite strict in case of a breach in the time schedule.

In fact, since bribery is hard to prove, activists have purposely chosen to concentrate on the legal violations and procedures surrounding the case. Farglory's contract states clearly that construction shall be completed within three years from the date in which the construction permit was obtained; the expiry date being June 30, 2014 (Arthur Yo, email communication, July 4, 2014). Since construction was not terminated by the set date and will continue for most of 2015, activists are of the opinion that Farglory has breached the contract. Furthermore, on June 11, 2014, even though the scandal about Farglory's bribes was still raging, the latest EIA (which began after Farglory had resubmitted the project with three ulterior changes) once more approved the Dome project "with conditions" (有條件通過 *you tiaojian tongguo*). The three changes dealt with the time schedule, the shifts and the number of coaches for the transport of the remaining raw material, needed for construction. In regards to the time schedule, in addition to the original 540 days forecasted, Farglory added 350 days of work, for a total of 890 days, forecated to complete the project (Lai, 2014a). Since the daily working shifts and the number of coaches had had to be reduced due to traffic congestion, night as well as weekend working shifts had to be increased, with an average of 40 coaches per hour during evening shifts (from 12 initially) to 34 per hour during the weekend and 48 per hour at night (previously eight), among rising discomfort for local residents. In regards to these recent changes, environmental activists and their lawyers claim that Farglory is blatantly violating several of the conditions initially set by the Taipei city government for this project. First of all the time limit for the transportation of the raw material, set in the contract, was March 15, 2014. Nevertheless, Farglory continued to transport soil and rocks (increasing working shifts) even after the deadline expired and the city government did not issue any violation ticket nor did it impose any fine on the company (Tsai Ya-ying, email communication, June 20, 2014).

Farglory has adduced as explanation that it can only dispose of the discarded soil by transporting it to Taipei's port of Bali (八里), filling in its embankments; however, the port's refusal to accept any more of the discarded soil material has considerably delayed the construction; for this reason Farglory requested an extension for the completion of the Dome project, exceeding the term originally written in the contract, which was June 30, 2014 (Lai, 2014a). The date was later pushed to mid-August 2014, and it was recently reported that Farglory proposed an additional delay to the project of 882 days as a result of slow progress. Since one of the reasons for wanting the Dome was to use it as the venue of the 2017

Taipei Summer Universiade, time is of the essence and the city government, one of the staunch supporters of the project, is facing great pressure, with activists calling on Mayor Hau Lung-bin to take a clear stand. Hau was quoted by the media, in the summer of 2014, reassuring the general public on the Dome's construction, stating that the completion date of the project will be no later than early 2016 (Lo, 2014). Pressured by social movements' concerns and by the recent bribery scandal and allegations, Taipei's former Mayor placed the project under a monthly progress review, supervised by high-ranking Taipei city government officials, to make sure that it will be completed for the Universiade. Hau declared that the company will be subjected to the severest punishment allowed by the law, should Farglory breach the contract and further delay the construction of the Taipei Dome (Arthur Yo, email communication, June 29, 2014).

Activists are furious at what they consider as complacent behavior on the part of the city authorities, ready to grant Farglory several leeways, in order to see the project completed. The anti-Dome movement has thus renewed its requests that the city government revert the facility into a green space for cultural activities.

Not all governmental agencies are keen to see this project completed. To name but a few, the *Control Yuan* (監察院 *jianchayuan*) and the *Agency Against Corruption* (廉政署 *lianzhengshu*) are among the most active organs in lodging complaints against the Taipei Dome Project. Notwithstanding their objections and even though protesters and activists managed to gather valuable informations to build a solid case against the Dome complex, the sporting facility is slowly but gradually taking shape, prompting many to fear that it is too late to stop it.[15]

Conclusions

Although my sources indicate that quite a few local residents were concerned about the possible impact of the project on their neighborhood, the truth is that the general public remained largely disinterested and indifferent in regards to the Taipei Dome Project. Unlike the Kuokuang case, there was no broad emotional or populist mobilization of any sort and the arguments put forth by environmental activists – the protection of green spaces – were not perceived as urgent concerns by many. This is partly due to the differences of opinion among the anti-Dome movement itself and more than anything it is largely an outcome of missing media attention and support in framing and popularizing the issue at stake; the consequence was a lack of resonance among the general population who were largely unaware or uninformed (and thereby indifferent) in regards to the construction of the Dome.

Environmentalists tried but failed to make people realize that the destruction of the factory grounds does not merely represent an ecological harm but is first and foremost the loss of an important cultural legacy, which is an integral part of Taipei's urban identity. In fact, the cultural heritage aspect embodies a focal point, with the Songshan Tobacco Factory representing the island's recent colonial past, and a cultural legacy of what makes Taipei's architecture so unique. There is an element of romanticism in preserving the cultural heritage of the bygone Taipei, incarnated by those trees that were planted on the side of the wide avenues of Guangfu South Road and Zhongxiao East road as far back as the early 1980s and are celebrated in the famous song by Wang Mong-ling (王夢麟), *Mumian Dao* (木棉道).

This case, like all other previously analyzed cases, is a perfect example to illustrate the depth to which environmental issues are connected to politics in Taiwan. The Taipei Dome was selected by Pan Han-sheng and by the Green Party to become a showcase for the legislative campaign in Xinyi District; Pan's reasoning was that should he have managed to obtain a positive result, it would have turned out as a good example of the Green Party's achievements.

Nevertheless, despite the solidity of the legal arguments put forth by activists and concerned lawyers, the city government, which has its own interests and stakes in the project, has so far managed to push the Dome project through, making the pro-development faction gain the upper hand in this urban renewal confrontation.

Notes

1 In response to the growth of Taipei and the increasing demand for office spaces, the military factory grounds of Xinyi (153 hectares) were chosen to become the city financial center in 1978. The Xinyi Development Project, under which the City Hall would be built and an international financial center constructed, was launched in 1987. For more information on the transformation and eastward expansion of Taipei see Jou (2005).

2 The Build-Operate-Transfer (BOT) model is a mechanism to get private capital to sponsor public infrastructure projects. More precisely, it is a form of project financing, where a private entity receives a concession from the private or public sector to finance, design, construct and operate a facility stated in the concession contract. This enables the project proponent to recover its investment, operating and maintenance expenses in the project in due time. The BOT model of financing has been used in almost every big infrastructure-project in Taiwan since the 1990s; a period in which the economy entered a phase of recession and public debt soared thus obliging authorities to seek alternative means of financing.

3 The "tree-hugging" phenomenon refers to cases in which individuals form protective cordons around trees scheduled to be transplanted or felled for developmental purposes. This trend has become increasingly strong in Taiwan and a Tree Party (樹黨 *shudang*) was established in August 2014 (becoming Taiwan's 259th political party) by Pan Han-sheng (former Green Party spokesman) and his brother Pan Han-chiang (潘翰疆), with the aim of fighting on behalf of creatures that have no votes and no voice such as trees and animals, rushing into the political campaign for the municipal elections of November 2014 (Lai, 2014d) in which the party successfully obtained Nantou County's Chichi Township mayoralty as well as a few other minor gains (Wen, 2014).

4 The 228 Massacre, also called 228 Incident by the Kuomintang, was an anti-KMT uprising that started on February 27, 1947 and was suppressed by the government of the Republic of China on February 28 (2/28), resulting in the death of numerous civilians. This incident marks the beginning of the White Terror era in Taiwan, which would last for four decades.

5 The Chipko movement is a non-violent movement started by Indian villagers in the 1970s and 1980s, to protest against the destruction of trees and forests, which were important resources that they depended on.

6 See for instance the controversy surrounding 32 old tress that had to be removed from the campus of Chiang-Tsui Junior High School (Lee, 2013b: 3), Even more recently, another controversy involving some old trees in Hua Guang has attracted widespread visibility in Taiwan; see, among others: Lai (2013a).

7 The reason why these auxiliary amenities and buildings have been added is that they offer a much better warranty that the developer can regain its costs and make a profit rather than if the construction would feature simply a dome for sporting events, which take place quite rarely in Taiwan.

8 Suspicions regarding corruption and bribes became even stronger when, on May 30, 2014, Farglory Land Development Co's (遠雄建設 *yuanxiong jianshe*) chairman, Chao Teng-hsiung (趙藤雄) was accused of offering bribes to former Taoyuan County deputy commissioner Yeh Shih-wen (葉世文) through middleman Tsai Jen-hui (蔡仁惠) – a retired professor – regarding a bid that the company had secured earlier in April 2014, for a government housing project. He was released on bail on May 31, 2014 (Shih, 2014).

9 The slogan plays with the words "搗蛋 *Daodan*" which literally means, "break the Egg" (a reference to the name of the sport stadium) but it has a double meaning since the phrase "*daodan*" is used in Chinese as the Western "trick or treat." In other words, "do our bidding or we'll trick you." (See Figure 5.7).

10 The Spirit of the Mountain Beast symbolizes the destructive attitude of men towards nature and the big loss (in quantitative terms) of green spaces during the developmental process of the Taipei basin.

11 Dafydd Fell, in his *Government and Politics in Taiwan*, points out that small parties in Taiwan are divided into two categories: purified and prophetic (Fell, 2012: 103). While purified parties are those basing their ideas on a pure version of ideology, which has been abandoned by bigger parties such as the KMT or the DPP, prophetic parties, among them the GP, base their claims and policies on completely new ideologies. Thus, GP and DPP have different backgrounds, different ideological claims and interests and, consequently, different supporters, all elements which made the Xinyi-Songshan District cooperation far from easy.

12 Deep Ecology is an environmental philosophy coined by Norwegian philosopher and alpinist Arne Naess in 1973. The philosophy purports the inherent, intrinsic value of all living beings, regardless of their utility for human needs and advocates for an immediate restructuring of modern human societies in accordance with such purposes.

13 The intersection between Zhongxiao East Road and Guangfu South is one of the busiest in Taipei and was so even before the sporting facility plus commercial and recreational buildings complex that will soon be completed. Therefore, in Arthur Yo's opinion, the initial choice of locating the facility here was a major mistake (Arthur Yo, interview, October 30, 2011).

14 In regards to the secretive and non-transparent behavior of the city authorities vis-à-vis this project, Ketty Chen, a concerned activist and scholar with a personal vested interest in this neighborhood (her parents are long-time residents of Xinyi), told me during an informal conversation that local residents were not notified about the future construction of the Dome until it was already being constructed (Ketty Chen, informal conversation, June 19, 2014, London, SOAS).

15 The election of a new city administration under Mayor Ko Wen-je in November 2014 has initiated a controversy between Taipei city and Farglory, regarding the future of the Dome. As of April 2015, both Taipei deputy mayors Charles Lin (林欽榮) and Teng Chia-chi (鄧家基) as well as Mayor Ko have been quoted by several media outlets saying that if the Taipei Dome fails the upcoming safety inspection, the city government will not rule out the possibility of terminating the construction contract with Farglory.

6 The Tamsui North Shore Road Project

Background information

The Tamsui area is known for several things but its beautiful mangrove (紅樹林 *hongshulin*) forests, gracefully decorating the side of the river, no doubt constitute one of the most famous tourist attractions. The Tamsui River runs beside the *Guandu Nature Park* (關渡自然公園 *guandu ziran gongyuan*). According to the historical records, the area was almost completely covered by water bodies at least until the early Qing Dynasty. In Emperor Guangxu's time (光緒帝 *Guanxu di*, 1871–1909), after the arrival of a few settlers, the area became farmland and remained so until 1965. In 1968 the government constructed an embankment for flood-prevention, while farmlands outside the embankment were gradually abandoned. The *Guandu Nature Park* has long been home to a variety of domestic and migratory wild birds. Due to its vicinity to the Taipei metropolitan area it has long been the target of interest by developers, attracted by potential high-density real estate projects (Tang and Tang, 1999: 361). Due to its geographical location, near the mouth of the Tamsui River, frequent tidal changes and the saline content of the water have transformed the location into a unique ecosystem, very suitable for mangroves, with a few heritage buildings and scenic areas. In the past decade Tamsui has become a popular attraction for daily visits from city slickers and families in search of a relaxing afternoon, away from Taipei's urban chaos. It goes without saying that the number of commuters between Tamsui and Taipei (from those who reach Tamsui for the weekend to enjoy its natural surroundings, to the growing number of local residents who go to Taipei for work) has been rising steadily. Additionally, Tamsui has seen an increase in Taipei residents moving to the area for its scenic beauty and lower housing and rental costs.

This incessant flow of people entering and exiting increases the demand for an upgrade of the available means of transportation. At the moment there are two ways to travel from Taipei City to Tamsui and vice versa: the metro line and a highway, *Route 2* (台二線 *tai er xian*), which is increasingly prone to traffic jams at rush hours, a condition I witnessed several times when living in Taipei. With the real estate industry building more and more skyscrapers to attract newcomers to Tamsui and increase their revenues, the existing means of transportation between the latter and Taipei City are becoming inadequate.

Transportation alternatives

The traffic problem dates back to at least the year 1996 even though the situation was not as serious as it is today; since then, numerous projects and proposals have been planned and discussed to solve the worsening traffic conditions; due to the many agents and interests involved, they were all either discarded or delayed until April 15, 2011 when the committee responsible for carrying out the EIA on this case authorized the construction of this controversial road. Environmentalists have fought this project since 1996 due to the serious consequences they fear it would have on the ecological environment and the rare species of birds and trees that live in it.

The four possible alternatives proposed during these 15 years by the various agents involved are: 1) the widening of Route 2; 2) the building of a highway along the north shore of the Tamsui River; 3) the building of a surface road along the north shore of the Tamsui River (on the riverbed); and 4) the building of a surface road along the north shore of the river (this time not on the riverbed). All of these projects present problems and have encountered fierce opposition by several of the concerned parties.

1 Widening of Route 2

The most viable option in the 1990s seemed to be the widening of the already existing *Route 2* (台二線拓展 *tai er xian tuozhan*). This alternative though was abandoned almost immediately, due to lack of funds. In fact, during the planning phase *Route 2* was designed to be 40 meters wide. Due to lack of subsidy to purchase the necessary land, its width was then reduced to 25 meters. The widening plan soon ran out of money and the whole project was eventually discarded.

2 Highway along the North Shore of the Tamsui River (1996–2000)

The second alternative (淡水河北側沿河快速道路 *danshuihe beice yanhe kuaisu daolu* – 高架 *gaojia*) was proposed by the Ministry of Transportation in 1996 and it obtained the formal approval of the Executive Yuan one year later, in 1997. The planned highway was set to be 12.8 kilometers long, linking the existing highway network of Taipei City to the northern tip of Taiwan island. The route started from Beitou (北投), proceeded towards the Guandu Bridge, crossed over the metro line and continued along the Tamsui River, passing through areas of natural beauty such as mangrove forests.

Shortly following the project's approval by the Executive Yuan, some civil society groups joined forces and set up an organization called the National Alliance for the Rescue of the Tamsui River (全民搶救淡水河行動聯盟 *quanmin qiangjiu danshuihe xingdong lianmeng*) in order to fight the Executive Yuan's decision. In voicing their objections, they demanded that a formal EIA be carried out. In 2000 the Environmental Protection Administration conducted its first EIA on the project. Once the EIA reached an end, EPA stopped the construction of the road, considering it too damaging for the local ecosystem.

3 Surface road along the north shore of the Tamsui River (on the riverbed)

The third alternative, which was to build a surface road on the north shore of the river (淡水河北側平面道路 *danshuihe beice pingmian daolu*) on the path of the actual riverbed (河川行水區 *hechuan xingshuiqu*), was put forward by the Taipei County government, but due to the various restrictions of the *Water Resources Law* (水利法 *shuili fa*) this project was discarded in the early planning phase. For the sake of clarity, a surface road is a standard road that runs at ground level. The term is usually employed to differentiate a particular road from a highway in that it usually runs more slowly as it has many traffic lights.

4 Surface road along the north shore of the Tamsui River (2006–present)

To avoid the above-mentioned problem, the Water Resources Bureau (水利署 *shuili shu*) of Taipei County later suggested building the expressway along the river (not over it) so as not to violate the *Water Resources Law*. This fourth proposal, titled *Surface Road along the North Shore of the Tamsui River* (淡水河北側平面道路 *danshuihe beice pingmian daolu*), is the one that best embodies the divergent interests of the various agents involved in the controversy: two local governments – Taipei City and Taipei County, or New Taipei City (新北市 *xin bei shi*) – the central government with its various bureaus and offices, the local community and various social groups.

The planned road in this fourth project starts from Deng Hui Avenue (登輝大道 *denghui dadao*), now called Danjing Road (淡金路 *danjing lu*), and it stretches along the river connecting the mangrove area with Zhu Wei Road (竹圍路 *zhuwei lu*). It measures 4.7 kilometers in length and it is a 20-meter-wide, four-lane road.

Environmental governance at play

As mentioned earlier, this last proposal, which was approved by the EIA committee in April 2011, best exemplifies the various conflicting interests of the many parties involved in the controversy, thus embodying a perfect example of the complicated mechanisms that characterize contemporary environmental governance in Taiwan. First of all, this case shows how different levels of the government can have differing interests in regards to a certain project. Specifically the *Tamsui North Shore Project* pits the New Taipei City government, which includes Tamsui Town and is in favor of the road, against the Taipei City government, opposed to the project. What are their respective interests? Taipei County, hoping to develop the real estate market in Tamsui, has pushed for the construction of the surface road for many years, while Taipei City, afraid of the traffic consequences for the already traffic-jammed roads, has long been staunchly opposed to it. At the initial stage, Taipei County, encouraged by the

Ministry of Transportation, sought hard to establish regular meetings with Taipei City in order to find a solution to the protracted stalemate. In 2006 various monthly assemblies were scheduled but Taipei City boycotted all of them, by refusing to send equal-level officials to the coordination assemblies (Lan Wei-gong, interview, May 12, 2011). In these cases, civil service personnel often create an informal network to facilitate coordination between local officials and their central counterparts; however, due to its opposition towards the project, Taipei City refused to send any representative of the city government to these meetings. Furthermore, at that time the central government was still in the hands of the DPP and between 2007 and 2008 the Minister of Environmental Protection was the already-mentioned Professor Winston Dang (陳重信 Chen Zhong-xin) who viewed the project negatively because of the damage it would cause to the mangrove forest (Winston Dang, interview, October 18, 2011). Although the movement for the protection of the Tamsui river area is long-standing, due to its richness in wildlife species such as migratory birds and precious mangroves forests, property values' appreciation in an area conveniently located right outside the greater Taipei region, has presented activists with mounting problems.

A story of mangroves in northern Taiwan

At the end of the 1970s, as the conservationists' current of Taiwan's early environmental movement was becoming stronger, the problem of the threat to the survival of the mangroves in northern Taiwan, due to unruly and unchecked development, became the *pièce de résistance* of conservationists. Experts and scholars tried to involve those few governmental agencies that were supportive of natural conservation into their struggle so as to give more strength to their fight against those advocating for development (Ho, 2006: 39). Conservationists' methods basically consisted in trying to gather support among governmental organs in favor of nature protection, while marginalizing the interests of those in favor of developmental projects. In 1978 the *Miscellaneous Plan for the Development of Taiwan* (台灣地區綜合開發計劃 *Taiwan diqu zonghe kaifa jihua*) by the Executive Yuan had already classified the mangroves in the Zhuwei area of Tamsui as a touristic area, which fell under the categories of protected zones. Later that year, the Taiwan Provincial Government,[1] together with the Taipei County Government, released a plan for the construction of public housing, in which it expressed the desire to have the marshes filled in and the mangroves forests cut down, so as to facilitate development and construction (Ho, 2006: 39). Even though mangroves were already protected by the previously mentioned plan, de facto numerous provisions for conservation were mere pro-forma regulations, which were not often consulted, let alone implemented. Further, mangroves protection is an issue that has long created frictions among different governmental departments. Conservation experts, conscious that mangroves protection was becoming a hot topic, tried to act as harmonizers between the different parties, on the one hand participating in governmental meetings on the issue and on the other explaining the importance of conservation to the masses. Due to growing national discussion on the mangroves issue, in 1980 the then president of the Executive Yuan, Sun Yun-suan (孫運璿) twice publicly spoke out in favor of the protection of the Zhuwei area forests. From that moment, the police started monitoring and punishing illegal behaviors that would wreak havoc on the forests. In 1983 the government established the *Guandu Waterbird Refuge*, which later became the *Guandu Nature Park*, with the explicit goal of helping to protect wildlife species of birds and plants, unique to the area.

Road chronicle

One of the main events in the road saga took place in July 2008, when the KMT was back in power after eight years. Wu Yu-sheng (吳育昇), an influential Kuomintang legislator, convened a meeting to discuss the fourth proposal in detail in order to get past the political impasse between the two local governments. He invited representatives from the Water Resources Bureau, from the Taipei County government, from the Transportation Bureau of Taipei City and from the Ministry of Construction (營建署 *yingjian shu*). The talks ended in deadlock mainly due to the opposition of Taipei City's Transportation Bureau; its representatives claimed that the road would mean a massive increase in the flow of people entering the city, which would be detrimental to Taipei's already jammed arteries.

What proved crucial for the temporary halting of the surface road option in the end was a prehistoric site that was discovered on the planned route in July of that same year, causing the central government to stop work and call in an expert from Taiwan University to evaluate the relics (Lan Wei-gong, interview, May 12, 2011).

In September 2008 a few social groups opposed to the construction of the project organized a press conference and started an online petition (反淡北道路連署 *fan danbeidaolu lianshu*) (Petition, 2008) broadcasting the fact that the consultation with the county government, organized by legislator Wu, did not yield any results. In the months that followed, a wide range of civil society groups joined in the collective effort to stop the project. The main group was the Tamsui Heritage Foundation (淡水文化基金會 *danshui wenhua jijinhui*), established in 1995 by the Tamsui municipality government to supervise the area's numerous cultural relics, which undertook the action of writing to EPA to remind it that the previous EIA was already eight years old and that, at the time when it was carried out, it had resulted in a "red light" for the project. In reality, the first EIA, conducted in 2000, was directed at the second project but since the planned route in the fourth project does not differ much from it, except for being shorter, some of the agents involved (e.g., EPA) thought it unnecessary to carry out a second EIA.

At the same time, this same group of NGOs, sometimes collectively referred to as the Alliance of Green and Democratic NGOs (綠色民主社會聯盟 *lüse minzhu shehui lianmeng*) prompted the Control Yuan to pressure EPA to renew the EIA while criticizing the pro-business mentality of the Taipei County government. Many of the social groups interviewed for this case have used the word "hoodlum" (流氓, *liumang*) to describe construction companies' personnel and local governments who, in their words, often utilize intimidating and rogue techniques to make them desist in their quest to stop several developmental projects.

The first turning point in the Tamsui Road case was October 2008 when EPA, set under pressure by the Alliance, sent a letter to the Taipei County Water Resources Bureau requesting that a new EIA be carried out, even though the proposed route was almost identical to that from the year 2000. Among the different civic organizations comprising the Alliance we find: the Green Citizens' Action Alliance, the Society of Wilderness and the previously mentioned Tamsui Heritage Foundation. By forming this alliance, green civil society organizations and concerned local residents have gained considerable strength in influencing a government organ (EPA), getting it to comply with their requests (namely, conducting a second EIA). While a few of these groups were obviously driven by private interests – some local residents were afraid that their properties would lose value by being so close to a major highway – most had more altruistic motives: some wished to preserve this area for recreational purposes, while others hoped to save the local wildlife and flora.

The second essential turning point took place in November 2008, one month after EPA ordered a second EIA, when the mayor of Tamsui Town, Tsai Ye-wei (蔡業偉) petitioned the Control Yuan to impeach the then director of EPA, Shen Shu-hung, claiming that Shen had violated the EIA Act. The grounds for the accusation were that the planned route was in fact only 4.7 kilometers long, and thus, according to EIA regulations, did not necessitate a new EIA. In this petition letter (陳訴書 *chensushu*) Tsai writes that EPA had simply "caved in to the emotional outbursts of 'pseudo-environmentalists' lacking any basic knowledge of the project."[2]

The director of EPA was in fact "temporarily" impeached, and the ROC's former vice-president (2008–2012), Vincent C. Siew (蕭萬長 Xiao Wan-zhang), declared in 2009 that the responsibility for planning this project was being returned to the Ministry of Transportation. Starting from that moment everything stopped as the different parties could not agree on a common solution. At the central level, EPA and the Ministry of Transportation stepped up their joint efforts to find a common solution while waiting for the results of the second EIA.

Different positions among governmental bureaus and pro-business faction

This case involves a plethora of different actors with conflicting interests and multiple levels of government, each with their own agenda. Even though the differing positions of the two local governments (that of Taipei City and that of Taipei County) have already been emphasized, it is necessary to briefly analyze the differing opinions and views present within the central government itself.

The Ministry of Transportation was the main organ in favor of approving the project in the hope that it would ease the difficult traffic situation. EPA, also a central government organ, initially claimed that an EIA was not needed, since the new project was less than five kilometers in length. According to Articles 5 and 6 of the sub-regulations of the EIA Act titled *Standards for Determining Specific Items and Scope of Environmental Impact Assessments for Developmental Activities*, any project that is less than five kilometers in length does not need to undergo an EIA. As previously mentioned, under pressure from civil society groups, EPA agreed that a new EIA was nevertheless needed.

The positions of the local governments are also worth analyzing in depth: on one side, both the government of New Taipei City and that of Tamsui Town are in favor of the new road and claim that the decision-making process rests with them since the road would run within New Taipei City. Thus, the right to build and operate this road (without consulting either the central government or the Taipei City government) is theirs. According to Article 11 of the Highway Law (公路法 *gonglu fa*), the right to build and operate a road within the territory of a county government resides with the latter.

Opposed to the road and worried that another flow of people entering the city would increase traffic congestion, the Taipei City Government's position is that it shall support the project only if it passes the second EIA assessment. It has to be mentioned that the Taipei–Tamsui surface road has long been planned as a lynchpin to New Taipei City. Since the very beginning though, the pressure to build and the support behind this road has come from New Taipei City and Tamsui Town, with Taipei City's stance on this issue varying from disinterest to outright opposition, according to a multitude of factors, among them: who was in charge of EPA; who was in charge of Taipei City's Traffic Bureau; and, finally, which party was in charge of the central government, blue (KMT) or green (DPP). All of our informants in the Taipei City government claimed that the city has nothing to gain from this road and thus has no interest in supporting it; according to them, there is sufficient communication between the two places, and an additional road is only going to increase traffic jams in downtown Taipei.

But what are the real motives of the "pro-growth" coalition aggressively advocating for the road? According to Wang Chung-ming, Green Party candidate for the position of City Councillor of Tamsui Town in 2010 and in the 2014 municipal elections,[3] and member of the Green Party Central Committee, in this case the political interests of politicians overlap with the financial incentives of the business community; in fact, quite a few of the local politicians also own successful real estate and construction companies, which hope for more infrastructure and housing building (Wang Chung-ming, interview, May 4, 2011).

In this regard, it is interesting to dissect the fallouts of the Fukushima incident and its significance for the Tamsui Road Project. In April 2011 several local leaders, especially those of Shimen and Wanli, expressed a favorable opinion towards the road construction, which they heralded as an additional route of escape, in case of a nuclear catastrophe. Legislator Wu Yu-sheng was quoted

Figure 6.1 Wang Chung-ming at a "No Car Day" protest in Tamsui, September 17, 2011 (source: Simona Grano).

saying that in case the road is not constructed the effects on the evacuation plan for NPP-1/2 would be scary (Liao, 2011).[4]

In democratic countries, politicians compete for votes. It thus goes without saying that the driving force behind several decisions regarding developmental projects is often the attempt to attract the electorate (Weller, 2006: 150). In the Tamsui case some members of the Legislative Yuan were opposed to the construction of the road (a notable example being Tian Qiu-jin, 田秋堇, a DPP legislator) while others, for instance the already-mentioned legislator Wu Yu-sheng, supported the project in order to fulfill the promises made during the election campaign, which was to relieve traffic congestion in Tamsui. During the past three years, various organs have tried to harmonize the positions of the various actors involved in this protracted stalemate, so as to facilitate the adoption of a common project. Nevertheless, the more active stakeholders in trying to coordinate and push forth the construction were also those who would gain the most from the project (e.g., Tamsui Town and legislator Wu; New Taipei City and the real estate sector). All of these agents, and especially the ones connected with the government, be it the local or the central one, know how to employ and bend the rules to their own advantage and try to influence the decision-making process. In the next section, the discussion will move on to the analysis of the methods and strategies used by environmentalists to fight the project.

Environmental activists' unofficial methods of contention

Many environmental groups have been involved in this case throughout the years; some are still active today and have meanwhile devoted their interests to other cases while other, less stable groups disappeared when the immediate threat to them or to their properties seemed to be more distant, as was the case in the early phases of Taiwan's environmental movement, with several local anti-pollution protests (Hsiao, 1999).

Since 2006, when the discussion about the fourth project began, a few of the more active social groups have decided to join efforts with the local residents of the Zhu Wei area. Being the closest area to the mangroves, its residents are those who will lose the most once the road is built, claiming that it will ruin their view and affect the local environment negatively (Chen Fang-bin, interview, May 3, 2011).[5]

In June 2008 a group of Zhu Wei residents called the *Green Citizens' Action Alliance* to enlist their help. For the sake of clarity it must be explained that the first big wave of protest (抗議 *kangyi*) against this project took place from 1997 to 2000. After carrying out the first EIA, EPA ended up halting the project altogether. The second time around (from 2006 up to the present), NGOs lament the fact that local citizens are much less involved and interested in fighting the authorities although social movements are unclear on why people have lost interest in the case (Tsui Shu-hsin, interview, March 24, 2011).

As for the second EIA, the Tamsui Road Project went through three review meetings in the year 2011 alone; during these assemblies, civil groups were

Figure 6.2 Tsui Shu-hsin and the author at GCAA's offices, March 24, 2011, Taipei (source: Simona Grano).

allowed to express their concerns and present their evidence while experts from both sides voiced their opinions and proposed alternative plans. Both sides had the right to call in three experts to champion their cause (Lai Rong-xiao, interview, September 1, 2011). Tsui Su-hsin explained how difficult it was for green groups to participate at such meetings: "I participated but I was not invited. Social groups are never welcome because government officials fear they might cause embarrassing scenes" (Tsui Shu-hsin, September 20, 2011). It goes without saying that to circumnavigate this obstructionist behavior on the part of the authorities social activists had to spend several hours monitoring webpages and online forums to find out, in a timely manner, when the EIA committee was going to hold its meetings; this is easier said than done since the committee purposely posts such information online less than one week prior to the actual meeting, in order to try to restrict social groups from participating.

In this case, environmentalists represented what Tang and Tang have defined as "diffuse interests" (1999: 363). In such cases the most suitable methodology to gain support is the employment of face-to-face mobilization techniques ranging from organized sit-ins in front of the Control Yuan with a crowd of 60–70 people (in 2008) to systematic street protests, generally held the day before the committee met. Furthermore, prior to every EIA assembly, social groups prepared a summary of the case, containing the most important facts, and they sent it to various media outlets, trying to capture their attention. This has ensured media coverage – even on short notice – but it has also meant extra work for the environmentalists, who have had to prepare press releases before every protest or demonstration in order to get the media interested. However, after Fukushima the media has become more interested in GCAA's actions and their anti-nuclear positioning. In fact, GCAA's two main issues of concern are fighting against the construction of nuclear plants on Taiwanese soil, as seen in Chapter 3, and protecting the Tamsui area (Tsui Shu-hsin, interview, March 24, 2011).

As shown with the three previous case studies analyzed, apart from media involvement, the most important factor for success in a confrontation with the government is public opinion, which is dependent on the degree of coverage by the media. In fact, while public opinion was the main factor creating pressure for the government in a pre-electoral year in the case of the Kuokuang petrochemical complex, it is precisely the overall lack of interest by the general public that made opposition in the Taipei Dome fail. As stated by Williams and Chang (2008: 134): "Successful environmental protection requires the direct participation and support of the public, more so than almost any other government initiative." Thus, whichever actor manages to win over public opinion has won half of the battle. All social groups interviewed in this case have stressed that public opinion plays a very important role in Taipei, where the political powerhouses have to gain the confidence of a wealthy, middle-class electorate (and where land prices are so high that every case involves huge economic interests), while it plays a lesser role in other Taiwanese cities (Song Hong-yi, interview, March 17, 2011; Tsui Shu-hsin, interview, March 24, 2011). While print media often report on green campaigns because they seek the catastrophist tinge that creates public interest and concern (and thus helps to sell more copies), NGOs have to be more creative in their dealings with the public. An effective way to keep people's attention high is to organize leisure activities, like concerts with famous artists and rock stars who perform on behalf of the protection of the Tamsui area, for example the already mentioned *No Nuke Street Band* (Yang, 2011), as well as coordinating other outdoor events such as biking trips close to the mangrove forests. These activities, meant to reconnect local residents with their natural surroundings, are aimed at portraying the negative impact that the construction of the road would have on people's lives and those of their loved ones. The message is clear: no more leisurely Sunday afternoons with children and grandchildren in the nice green surroundings of Tamsui. If the road gets built, the mangroves will be torn away and so will the bike lane! Finally, the organization of petitions was also quite successful in the Tamsui case with more than 10,000 signatures being collected in 2008 alone, the majority though from people who were not local Tamsui residents.

Political allies

In light of the common position vis-à-vis the project, social groups have often tried, in the course of the past few years, to find meaningful ways of cooperation with the Taipei City government. Gradually, they have come to realize that the Taipei City government does not want to publicly express support for NGOs and in those rare occasions in which it does help, it secretly leaks information on the misbehaviors of either the Taipei County government or the Tamsui Town government. Before the ruling in April 2011, social groups were uncertain about the results of the second EIA: "There is a 50–50 chance that things will go either way" (Song Hong-yi, interview, March 17, 2011; Tsui Shu-hsin, interview, March 24, 2011; Wang Chung-ming, interview, September 17, 2011). After the presidential election of 2008, in which the KMT regained the presidency, social groups became more pessimistic in regards to the outcome of the second EIA. In fact, the predominant opinion was that even though the City Government remains staunchly opposed to the construction of the road, its administration still belongs to the KMT and thus the central level might try to influence its stance, harmonizing it with that of the local Tamsui government. Even though the DPP government was also keen on doing business at the expense of the environment, social activists nevertheless maintain that it would have been easier to find a wedge between the different parties, to be used as leverage in their favor, when they were not all "on the same side" (Tsui Shu-hsin, interview, September 24, 2011).

As seen throughout the first chapters of the book, the initial trust that existed between the DPP and green groups is long gone and hard-core environmental activists – and in particular those belonging to the newer generation – are increasingly skeptical in regards to the DPP's environmental commitment. Thus, the only viable option for cooperation between environmentalists and politicians seems to be the Green Party. Wang Chung-ming, the already-mentioned local resident who ran for the post of City Councillor in Tamsui Town in 2010, representing the Green Party, was able to collect 7,098 votes despite his firm opposition to the construction project. The result is a success considering that his position ran counter to the wish of the majority of local residents who were not opposed to the road project; unfortunately for him, a city council member usually needs around 15,000 votes in order to be elected. Nevertheless, social groups view his performance as positive since his candidacy did not end up as a

Figure 6.3 "No Car Day" organized by the Tamsui Town government, September 17, 2011, Tamsui (source: Simona Grano).

Figure 6.4 Counter Parade organized by green groups, September 17, 2011, Tamsui (source: Simona Grano).

Figure 6.5 Counter Parade organized by green groups, September 17, 2011, Tamsui (source: Simona Grano).

political debacle and take this result as an indication that the number of local residents who are opposed to the road is in fact increasing (Tsui, 2009).[6] Again, considering the year in which the second wave of this protest was taking place, social groups – and especially the Green Party – have sought to exploit the Tamsui North Shore Road Project, turning it into an important campaign issue in the pre-electoral year of 2011. On September 17, 2011, on the occasion of a "No Car Day" celebration (organized by the local Tamsui Town government to boost its image of an "eco-friendly" administration, green groups, local activists and green politicians organized a counter-protest to emphasize the paradox and absurdity of a "No Car Day" when, upon completion of the project, the traffic flow will in fact be much worse than it is today (Tsui Shu-hsin, interview, September 17, 2011; Wang Chung-ming, personal communication, September 17, 2011). Alongside they also tried to attract votes for the Green Party and spread awareness in regards to the hypocritical attitude of the local government.

Rules-based participatory approaches and latest developments

While the road project was initially scheduled for completion by 2016, social movements – convinced of the negative impact the road would cause on the coastal area, on a precious wetland and on the mangrove forest – managed to stall the project, through the employment of rules-based participatory approaches.

In fact, the case against the road was formally terminated on April 15, 2011 when the third EIA review meeting ruled that the Tamsui North Shore Road Project could proceed. The experts responsible for the EIA named a few conditions that New Taipei City needed to fulfill before starting construction. The first condition was that New Taipei City would have to carry out a study on how to best conserve local cultural and ecological properties, by setting up an ad hoc committee for ecological supervision. Second, such environmental monitoring had to last for five years. Third, New Taipei City was expected to organize a workshop for the ecological supervision of the area; this workshop had to include at least 12 members recruited from among state officials, NGOs and the academic community in equal proportions (Central News Agency, 2011). Most activists interviewed considered these provisions as a mere façade, put up on the part of the EIA committee, to stifle criticism from the environmental community (Wang Chung-ming, interview, May 4, 2011; Tsui Shu-hsin, interview, September 17, 2011, among others). Under the current structure, with six members recruited among pro-KMT professors and experts, environmentalists maintain that, quite often, the committee will lean towards the approval of controversial developmental projects (Winston Dang, informal communication, August 31, 2011). As for the Taipei City government, initially opposed to the road, after the committee's ruling, it said that it would comply with the decision, provided that the road would stop before entering the city (Lan Wei-gong, email communication, July 12, 2011). According to our informants in the environmental camp, this type of behavior is a "somewhat childish attempt at not losing face, while being forced into complying with the EIA committee's decision" (Wang Chung-ming, interview, May 4, 2011; Lai Rong-xiao, interview, September 1, 2011; Pan Han-sheng, interview, September 1, 2011).

Neither social groups nor the Green Party participated in the last EIA meeting but they knew that the committee was going to meet on April 15. For this reason, they called a press conference the day before the ruling, to give visibility to the Tamsui Road saga once again, hoping that this would somehow influence the EIA committee's decision. As Wang Chung-ming put it: "We thought that the committee responsible for the EIA would behave reasonably" (Wang Chung-ming, June 29, 2011). What he meant was that, legally, the government should not have gone ahead with the approval, since the project violates the *Environmental Protection Law*. Strictly speaking, though, this claim is incorrect as the law states that if the road is shorter than five kilometers (here the road will measure 4.7 kilometers), it can go ahead with the construction without even being subjected to an EIA, which in this case was mandated anyway due to social groups' pressure on EPA. Furthermore, those in favor of the road maintain that it will have no effect whatsoever on the mangroves since no portion of the forest will have to be removed because the road will simply pass over it, without affecting the mangroves. Social groups refute this explanation and claim that flora and fauna will nevertheless be disturbed by the close proximity to a highway. Because social groups know that their leverage in cases like these (involving many powerful agents) is quite limited, one of the main strategies employed was to postpone the approval of the case. In an attempt to gain more time, they tried to sway public opinion on environmental issues by organizing numerous press conferences in order to put pressure on the members of the EIA committee. GCAA also sought to make use of the *Wetland Conservation Law* (溼地保育法, *shidi baoyufa*) to protect the mangroves area but since in 2011 the law had not yet been approved (it was in 2013), they were told by governmental officials that there was no legal violation since there was no law in place for the protection of the mangroves wetland (Tsui Shu-hsin, interview, September 17, 2011).

Even though the EIA committee's decision did not surprise those opposed to the road, the speed at which permission was granted is somewhat puzzling. On May 4, 2011 Wang Chung-ming and Chen Fang-bin,[7] a resident of Zhu Wei nicknamed "Chocolate Daddy" (巧克力爸爸 *qiaokeli baba*) unveiled, in an interview with the author, what they planned to do next.

Initially in May 2011, aggrieved social groups and local residents planned to sue the New Taipei City government (Liao, 2011); they claimed that one portion of the planned road, close to the mangroves area, slightly invades the boundaries set for the protection of the forest, thus violating both the *Cultural Heritage Preservation Act* (文化資產保存法 *wenhua zichan baocun fa*) and the *Environmental Protection Law* (環境保護法 *huanjing baohu fa*), specifically article 20, endangering the delicate ecosystem. The Taipei County government did not seem to be too concerned about the plan to sue and even retorted that social groups were wrong and that they simply used a different kind of measuring system when in reality, according to the County government's own calculation, the planned road was never to cross the forest. The plan to take legal action was abandoned, in July of that same year, for fear that it would require too much effort and money to sue the local government (Wang Chung-ming, interview, June 29, 2011).

Nevertheless, social movements continued to battle the road. Emphasizing the environmental damages the road would create, basing themselves on article 8 of the EIA Act, activists urged EPA to let this case enter a second stage of review (Pan, 2013: 1); in August 2011 they filed a complaint with the Appeal Commission of the Executive Yuan, asking for a repeal of the EIA results. This request for appeal was rejected by the Executive Yuan's *Committee for Petitions and Appeals* in February 2012. Therefore, Wang Chung-ming and the Green Party filed a lawsuit with the High Administrative Court of Taipei soon after, precisely on April 17, 2012. The court took more than a year and a half to carry out an inspection and to deliberate. On September 4, 2013 it ruled in favor of environmental activists by declaring the EIA report, carried out by EPA for the proposed expressway, as invalid on the grounds that EPA had discarded the damage to the mangroves even though, in certain parts, the road would actually encroach upon the forest (Lai, 2013b).

Even though EPA responded that it would study the Court's decision and then appeal, the ruling is undoubtedly a major victory for environmentalists. The New Taipei City government and the developers, on the other hand, have reacted badly to the High Court's decision; since actual construction started more than two years ago (in August 2012) they have quantifiable monetary and financial losses to deal with, now that the project lies, once again, dormant. While the Taipei City government has said that it will not begin with the construction on its side of the road until the legal dispute is settled definitively, the New Taipei City government has claimed that it will continue and prepare all relevant documents to submit to EPA for a second stage of the EIA process (Lai, 2013b; Wang Chung-ming, email communication, October 12, 2013).

The novel figure of the lawyer/activist, which plays a pivotal role in all of the controversies analyzed throughout the book (as well as in almost all recent environmental disputes of a certain visibility) is incarnated here by two young barristers, Zhang Yu-yin (張譽尹) and Huang Zheng-qi (黃正琪), who have been so kind as to consent being interviewed by the author.[8] The two of them have patiently explained to me the legal strategy adopted in pursuing this case. They maintain that the North Shore Surface Road will destroy the mangroves reserve, which has been designated as a protected area, according to the *Cultural Heritage Preservation Act*, which clearly states that it is forbidden to alter or damage the natural state of such areas.

In lawyer Zhang Yu-yin's words:

> This road, with its steady flow of cars, day and night, with its smog, its pollution, and its vibrations will certainly damage the nature conservation area. Presently, we have not yet legally sued the developer and the New Taipei City government; for moment, in light of the victory in our favour,

we maintain the New Taipei City government needs to stop construction immediately.

(Zhang Yu-yin, email communication, August 10, 2014)

However, in the area outside the mangroves nature reserve zone, work is under way.

Zhang Yu-yin has stated that in case the Taipei County government does not stop construction, they will have no choice but to take disciplinary actions, by requesting the Executive Yuan to rule in favor of a temporary halting of the road.

Furthermore, Zhang and Huang have explained how a major issue of contention regards the controversy surrounding land expropriation – itself a sensitive topic in contemporary Taiwan, in light of the numerous, disputed, land expropriations that have taken place in the past years, for state-led or private-led developmental projects.[9] In the case of Tamsui, since the road project actually impacts on the mangroves forest, the Taipei City government has not given its consent to carry out any land expropriation (particularly of certain agricultural areas for which disputes with current landowners could arise). In fact, the land close to the natural reserve has not yet been expropriated. The land on which the road should be built involves several different owners: a section of it is public land, which belongs to the state but there are several smaller parcels that belong to private owners; concerning these privately owned parcels, the requisition has to be carried out according to the relevant articles of the *Land Expropriation Act*. Currently, apart from a small number of private owners, who have already sold their land to the New Taipei City government at the agreed price, the majority of the land is still in the hands of these private landowners.

The anti-Tamsui Road movement believes that, in light of the uncertainty surrounding the future of the North Shore Road Project after the nullification of the EIA in the summer of 2013, the New Taipei City government should stop all land negotiations; on the contrary, the New Taipei City government is actively carrying out discussions on land expropriation and related matters, prompting the anti-Tamsui Road movement to dub its behavior as either "reckless" or "sure about the results of the appeal" since it is acting as if the EIA had not been declared invalid (Zhang Yu-yin, email communication, August 15, 2014). In fact, even though the second examination has not been concluded yet, the New Taipei City government has proposed to carry out land expropriation on certain lands that would be needed for the project; the problem lies in the fact that in case the road is eventually canceled, then those affected landowners will have given up their lands in vain (Zhang Yu-yin, email communication, August 10, 2014).

Furthermore, the two lawyers have devised a few legal offenses in the behavior of the New Taipei City government; in fact, among the 5.79 hectares that will have to be expropriated for the project, 5.74 hectares (constituting 99.19 percent of the total surface) are categorized as "land for special agricultural usage"; in the regulations set up for land expropriation, there are some specific provisions in place that clearly state that land which is destined for agricultural or pastoral usage should not be expropriated, with the exception of a very small percentage in case it is impossible to avoid doing so (Huang Zheng-qi, email communication, August 11, 2014). Social activists are keen to emphasize that drawing up an expropriation plan for parcels of land that had originally been destined specifically for agricultural use would require a cautious assessment as well as the involvement of several relevant bureaus, such as the Transportation Bureau and the Water Conservancy Bureau among others.

As noted by social movements, this is not the first time that EPA's EIA process results have been flawed, suffered a setback or been upturned by the Executive Yuan or by some other judiciary organ, clearly illustrating that, in its current form, the EIA is fraught with problems. EIA reviews, according to several of the lawyers who fight on behalf of the environment, interviewed for this book, often end up in an endless chaos; thus, substantial changes and transparency are urgently needed, which would allow for better following of the contents and procedural steps of the most important passages of the EIA process before a green light is given to controversial projects. According to Chan Shun-kuei, the responsibility for this case being rejected rests with EPA (Chan Shun-kuei, email communication, January 23, 2014); therefore the involved lawyers ask the New Taipei City government to abandon its idea of appealing the verdict and urge the agency to carry out a serious self-examination of its own role; for the future, they hope that more public participation in EIA meetings and proceedings will be allowed (Wang Chung-ming, Facebook post, August 13, 2014). At any rate, since the High Court's ruling of 2013 the case has lain dormant for more than a year.

Differing opinions in regards to the Tamsui North Shore Road Project

While New Taipei City Mayor, Eric Chu (incumbent since December 2010), maintains that the road has to be built in order to solve the difficult traffic situation affecting the tract of road between Zhuwei and Hongshulin, lawyer Zhang and the anti-Tamsui Road movement claim that the main culprit for the several delays and for traffic congestion is the protracted construction of several sewage pipelines, which keeps getting delayed (Lai, 2013b; Zhang Yu-yin, email communication, August 10, 2014).

Lawyers as well as social activists believe that, in order to solve traffic congestion, there are several methods that could successfully be employed. The first one would be the widening of Route 2, especially in its more narrow points; second, traffic congestion alleviation could be achieved by building the Tamkang (淡江 *danjiang*) bridge between the two banks of the Tamkang River. According to the anti-Tamsui Road movement, the current solution of building the North Shore Surface road will merely manage to "migrate" the traffic problem from Zhuwei to Taipei City's Dadu Road, which is where the new road would terminate.

Chocolate Daddy furthermore believes that many people in Tamsui are actually opposed to the construction of the road but do not dare to say so, due to the widespread connection between local political factions (地方派系 *difang paixi*), construction companies, and real estate developers and their influence at the localities, which, according to him, dominates local politics in Tamsui.

He shared with the author his opinion that many Tamsui officials have their own real estate and construction companies and thus have a vested interest in this road being built. He further claimed that local mafia (in accordance with construction companies) stages most of the traffic congestion problems on Route 2 in order to give people the idea that the road has become inadequate for the increase of traffic, saying that every time he goes *in loco* to check on the traffic situation, there is some staged problem like a broken light or a temporary blockage set up in order to slow down the traffic (Chen Fang-bin, interview, May 4, 2011). The fact that there are no workers actually repairing the alleged malfunctions has led him to suspect that it may all be a ploy to garner popular support for building an alternative road. When we went along with Green Party representatives to check on an "electrical problem" with a traffic light, on March 20, 2011, we were quickly shooed away by a bunch of workers who, previous to our arrival, seemed to be very engrossed in a card game rather than in fixing the problem. While I have no other proof than my own observations and what was told to me by my interviewees to support such claims, my feeling is that their allegations are not unfounded; after all it wouldn't be the first time that developers have hired private security companies to intimidate protesters and activists in order to carry out their projects undisturbed (Cole, 2013).

Conclusions

The Tamsui North Shore Road Project is a perfect case to illustrate the different levels of authority coming into play in the environmental governance process of contemporary Taiwan. Like all the other previously analyzed cases, the Tamsui Project was mostly driven by ideological and political motivations and exploited and manipulated to attract votes, by all the politicians involved, whether opposed or in favor of the road.

In this case, various social interests pull the government in different directions, as the gap between environmental concerns and business interests grows wider. Here, a vertical power hierarchy is clearly visible with social activists (from below) trying to counter powers from above (namely the local government) by using official channels of contention (legal appeals) and environmental policies (EIA) to aid their cause. In this specific case, though, we have also seen other powers at work: the conflicting claims of supervision for the project on the part of the several local governments involved (and their various bureaus) and the interests of both some local residents and civil society organizations in protecting the mangrove areas and or/their personal interests, such as property values. Under popularly elected mechanisms, different governmental bureaus all share the common interest of attracting votes. It was therefore not difficult to predict that in the end the outcome of the

EIA review would satisfy the interests of the majority of the residents of Tamsui, charging ahead with the development of the transportation system. What is surprising though, is that activists' endurance and apt employment of the available rules-based participatory channels yielded a powerful victory in 2013, with the ruling in their favor, which upturned the previous decision by EPA.

Thus, this case is a perfect example of how governmental action can be put under pressure by both unofficial tools (media, public involvement, street protests) as well as official ones (the already-mentioned legal channels), with the ultimate effect of delaying the project. Nevertheless, even though both unofficial strategies (protests in front of various governmental bureaus, petitions and media involvement) and official strategies (making use of the EIA regulations) were utilized in the Tamsui North Shore Road Project, the official strategy (filing an appeal at Taipei's High Court) certainly proved more effective, even if the road continuation is not yet ruled out.

It is possible that the transverse alliance of social groups with the Taipei City government partially aided civic associations and helped in delaying the final decision for a longer period of time, thereby proving that although environmental policies are, for the most part, mandated from the top, at the local level their implementation can be bypassed, altered or stalled by these various agents. Most importantly, the Tamsui North Shore Road case demonstrates that governmental bureaus and agencies at different levels can have different agendas and do not always harmoniously work together. Generally speaking, "opposition to" or "promotion of" a certain project depends on the specific details and characteristics of that project and of that locality. In this case the Taipei City government is of the opinion that the road project would negatively impact on the traffic situation and the environmentalists believe that the road would damage the local environment. In certain cases, a bureaucracy can experience internal conflicts over certain aspects of any given project. Here, for example, the central government's ministries were not all on the same page, with some agencies favoring the project – the Ministry of Transportation – and some against it – the Environmental Protection Administration, whose attitude changed throughout the course of the debate.

Notes

1 Since 1997 most of the Taiwan provincial government's functions have been transferred to the central government of the Republic of China following a constitutional change; currently the Taiwan provincial government effectively becomes a nominal institution under the administration of the Executive Yuan.
2 On file with the author. Courtesy of Professor Tang Ching-ping.
3 Taiwan's 2014 municipal elections saw the Green Party of Taiwan obtaining a victory with two elected offices won by Haoyu Wang (王浩宇) in Taoyuan County and by Jay Chou (周江杰) in Hsinchu County. See more at: www.greenparty.org.tw/news/20141130/216#sthash.nc75otDi.dpuf (last accessed on February 10, 2015).
4 On this point the secretary-general of GCAA Tsui Shu-hsin notes that the decommissioning year for NPP-1 is 2018 and thus to avoid any nuclear catastrophe the government should just keep true to its plans for NPP-1 ceasing to be operative (Tsui Shu-hsin, interview, September 20, 2011).

5 It is also worth mentioning that apart from Zhu Wei residents, the majority of Tamsui Town locals are in favor of the construction project or at least do not dare say otherwise openly (interviews with various local residents, February–May 2011).

6 His electoral success actually needs to be credited to his personal charisma. In fact, Wang Chung-ming, current spokesperson of the Green Party, is something of a small celebrity in Taipei: very active on social media he constantly updates his latest environmental stunts on his Facebook account, which is avidly followed by more than 3,000 people (including the author).

7 Mr. Chen is a well-known figure in Zhu Wei both for his political activism in the Tamsui Road case and because he owns a stall where he makes chocolate waffles, hence his nickname. For personal safety reasons, he has asked me not to quote his real name so I shall use the pseudonym of Chen Fang-bin.

8 The interviews have taken place via email and skype, through a set of fixed questions, regarding the current legal strategies that their team is pursuing in the Tamsui Road controversy.

9 The two lawyers imply that transparency and rule-abiding behaviors are not the strong points of KMT officials, after all the human rights infringement cases carried out by KMT cadres while expropriating land such as Dapu, in Miaoli and the Wang Family case in Wenlin (文林) (Zhang Yu-yin, email communication, August 10, 2014; Huang Zheng-qi, email communication, August 11, 2014).

7 Towards a more balanced and inclusive environmental governance approach

> Generally, a country that does not respect human rights also does not respect the environment; then there are countries like Taiwan, which respect human rights – at least apparently – but not the environment.
>
> (Echo Lin,[1] Secretary-General of the Environmental Jurists Association, October 20, 2011)

If this comment might seem a bit harsh at first sight, I would like to recall what happened to a friend of mine, as she tried to bridge the gap between the American and Taiwanese environmental and academic communities.

When we first met in Taiwan, in June 2011, she was working as the personal assistant for the director of the Environmental Sustainability Program at National Taiwan Normal University (師範大學 *shifan daxue*), Dr. Yeh Shin-cheng. At the end of June, an American delegation from the University of California came to Taipei to swap experiences on how to best achieve sustainability in university campuses around the world. The meeting, where I was also present, took place at Linda Arrigo's office in the campus of Taiwan Medical University. After talking to all the relevant persons in charge of sustainability at various universities around Taiwan, the American visitors requested to meet some representatives from the environmental community. At this point my friend organized a meeting for them with Calvin Wen, of Taiwan Greens and Robin Winkler, founder of Wild at Heart. After he discovered his visitors' plans, Dr. Yeh, who acted as their liaison in Taiwan, gently but firmly had them told that the meeting had to be canceled, mainly because the Taiwanese government was paying for their trip, thus there might be some problems at higher echelons if they were to meet with activists and furthermore their meeting could cause trouble for him. At this point my friend had no choice but to obey her superior's orders, cancel the meeting and lie about the real motives behind the change of plans. While she was distraught, feeling that she should not have organized the meeting without previously asking her boss, the sad conclusion that we can draw is that even high-ranking governmental officials in charge of "green issues," such as Dr. Yeh, have a still "limited" commitment towards the environment, which is often restricted to those façade-like activities that do not threaten the government, such as recycling on campuses and tree-planting, rather than truly promoting environmental sustainability, which would entail freedom of information and association on part of all the involved stakeholders. Such "domineering" behaviors continue to affect the relationship between environmentalists and the government, characterized by a growing lack of trust towards authorities' true commitment vis-à-vis environmental concerns, often scapegoated for "greater" economical goals or in order to achieve and maintain ample power on activists.

The really troublesome aspect of this event is that a very subtle form of authoritarianism seems to be sneaking back into Taiwanese academia. It is yet not clear to me whether there is a surveillance apparatus that actually checks on academicians (and in this case their contacts) and reports on their activities, actions and meetings to higher levels of the governmental apparatus or whether academics employ a self-restraint or self-censorship attitude in view of their political aspirations (as is probably the case with Dr. Yeh). In any case, one thing seems to be strikingly clear, the future that awaits environmental protection and sustainability in Taiwan is far from positive.

In Linda Arrigo's words:

> If the director of the Commission (even though pro-tempore) for environmental sustainability of the Shida University, whose goal should be to protect environmental sustainability, rather than cooperating with social groups openly distrusts them, boycotts them and blocks any attempt at networking between them and external guests, where does that leave us on the topic of promoting environmentally friendly practices and, even more disturbing, what does it say about Taiwan's political democratization and transparency?
>
> (Linda Arrigo, informal communication, July 20, 2011)

Furthermore, while rules-based-participatory approaches have been growing in importance and availability, numerous problems continue to hamper the implementation of the framework in place to protect the environment. In fact, the EIA system has improved but is still flawed and often a mere "rubber stamp" to seal already settled deals. Proof of this can be found in the testimony of many experts and scholars involved in such processes and interviewed for this book. Above all, a story that Professor Wu Kuen-yuh told me during our interview conveys a vivid description of the authorities' – at times still non-transparent – conduct.

A few years ago, as the government carried out an Environmental Impact Assessment on its petrochemical industry policy (化學工業政策環評 *huaxue gongye zhengce huanping*) Professor Wu was summoned as an expert on the EIA committee. After taking part in a few meetings and getting acquainted with the material, he concluded that the plan to carry out that review was a mere sham. "The plan to build the eighth naphtha cracker and to continue with the expansion of the fifth phase of the sixth naphtha cracker was already established; the assessment was simply a rubber stamp to give the whole process credibility" (Wu Kuen-yu, interview, September 20, 2011).

Based on first-hand observations, this book has attempted to show how a newer generation of activists, through the use of both the legal system as well as available social media, information campaigns and mass mobilization techniques, has been able to transcend traditional pre-existing boundaries, which had so far prevented a cross-pollination of common interests among several groups, giving rise to a unified movement capable of challenging governmental authorities and successfully influencing state policies. Such results were made possible due to the strategic timing during which the protests analyzed took place, in the years preceding the 2012 national presidential election, with the government under pressure and wary of dissenting opinions. The fact that every single project examined has been met with varying degrees of resistance from the local population as well as national NGOs, which managed to either successfully stop or stall construction, seems to confirm Robert Weller's (2006) observation when he remarked that Taiwanese in the 1980s were finally "discovering nature," by slowly becoming aware of the pollution around them, which was relentlessly destroying the environment. In fact, this book has depicted the gradual transformation of Taiwan's environmental movement throughout the decades; from an early period after the lifting of martial law when street protests as a legitimate mean to express dissent erupted, through a "lethargic" phase in which the movement was quieter and more institutionalized, until the KMT's return to power, which has given rise to a newer wing of environmental activists, at the same time more radical as well as more technologically savvy and skilled at employing the participatory means at their disposal.

Throughout the years, Taiwan has become more balanced, with political and business interests occasionally forced to bid farewell to long-hoped-for developmental complexes, in light of citizens' active involvement in fighting projects they do not agree with. However, much remains to be done to ensure that these channels are employed in the most transparent and correct way for a sustainable economic growth, which respects the island's precious natural surroundings. An important step in this direction could be achieved by freeing those in charge of environmental protection from political interferences, so that they could truly commit to the welfare of the environment without merely paying lip-service to nature conservation, as in the above-mentioned case of Taiwan Normal University. In fact, it is often the case that legislators, under constant pressure by the industrial sector and developers, quickly yield to the wishes of the pro-development faction without so much as a second glance at the long-term well-being of natural resources. Such conniving behaviors on the part of the authorities have provoked a widespread sense of mistrust among the general population, which has resulted in the resurgence of social activism. The recent bout of civic activism, no matter the formal cause activists stand for, stems in fact from a profound unease towards Taiwan's democratic foundation, a dissatisfaction with the increasingly polarized party politics, ineffective representative democracy and the country's widening social inequality. This vigorous activism has partly resurfaced due to the DPP's betrayal of its earlier pro-environmental profile all throughout its national tenure. Its inconsistent behavior disappointed activists leading them to think of the KMT and the DPP as differing little in terms of their environmental commitment. This can partly explain why smaller parties, such as the Green Party and the Tree Party, among others, fared quite well in the municipal elections of November 2014, thus indicating that the general population's awareness of "lesser issues," such as environmental and animal protection, is rising. In turn, such feelings have united activists with different agendas, who have thus attained an unprecedented level of efficacy in reaching across the political divide, making their concerns widespread and less partisan; this is especially true for what concerns NPP-4 and the "victory" in halting the facility, at least for the time being. Thus the country has witnessed a more broad-based participation in environmental affairs, which crosses the highly charged boundaries of politics in Taiwan, uniting people of different political orientations, under the common banner of environmental protection.

The Fukushima incident first and the Sunflower Student Movement in the spring of 2014, which began as a reaction towards the future of the island and its position vis-à-vis the People's Republic of China, have been turning points in easing the conditions for the re-emergence of civil activism in a period that was already of resurgence for civil society since 2008: such efficacy was reached thanks to the peculiar merging of a host of domestic circumstances and events, favorable to the environmental movement, that have relayed the fallouts of an international incident such as Fukushima, at the domestic level (Ho, 2014a). This merger of domestic and international factors can help to explain why Taiwan ended up generating the strongest anti-nuclear reaction in East Asia, showing why additional study of Taiwan's environmental governance is needed.

Sustainable development requires public participation coupled with political, institutional and administrative capabilities to achieve effective policies. On the one hand, empowering the public by giving it more institutional channels to vent its discontent through policy mechanisms, can advance democratic governance and prevent embarrassing street protests that have often occurred in Taiwan's past and have resurfaced since 2008. On the other hand, too many actors can actually further complicate the situation and hinder an effective management and implementation of the law. Therefore, Taiwan needs to create new and better mechanisms that legitimate and guarantee the participation of all concerned stakeholders in the process of implementation, in order to strengthen agency-constituency communication channels in environmental management. At the same time though, policymakers should avoid creating a situation in which these multiple stakeholders lead to a paralysis of the decision-making process due to lack of clarity on who holds ultimate authority, on various interconnected sectors.

As we have seen, determining factors contributing to an effective environmental management situation include the level of environmental education of citizens, the availability of participatory channels allowing individuals to take part in environmental issues and processes, the income level of a specific locality – crucial to understand if locals will welcome or reject a project – and, most importantly, local leaders' attitudes towards the environment. The theoretical framework of this book has shown that all the case studies analyzed represent models of environmental governance in which there is no uniform attitude towards the construction of development projects but rather a complex interaction of the state, local communities, social activists and other actors, all with conflicting interests. Environmental politics in contemporary Taiwan have evolved into protracted and exhausting deadlocks, involving the use of environmental laws and personal ties to advance each faction's goals. Environmental governance has been influenced by numerous heterogeneous actors who gained conspicuous influence after the democratization process started, towards the end of the 1980s; while authoritarian legacies are still present at the localities, in the form of patronage and favoritism, other – previously powerless actors – now also play an important role. Thus, it is fair to say that Taiwanese environmental governance is, especially at the local level, fragmented and inefficient, subjected to several debates that often prevent the best solution from being found; however, at the same time, the commitment towards environmental protection on the part of different previously marginal actors, is getting stronger. In this kind of environmental governance model, the state (especially the local one) often seeks to enhance its own power by entering into partnerships with commercial and other actors, lacking an overt political function, in order to increase its efficiency while seeking support for its developmental projects. The growth-machine framework mentioned in the introductory chapter thus fits perfectly for all the cases studied in the book. In the anti-Kuokuang movement, it was the central government as well as the local one that actively sought and welcomed the developer's aid in industrializing this rural locality; the Dome case also fits the model perfectly, by representing an urban renewal case, in which the commercial interests of local authorities and business corporations unite and clash against environmental activists' wishes to preserve one of the few green spaces left in downtown Taipei; the Tamsui North Shore Road Project is once more an example of how local politicians' interests and electoral promises get intermingled and advanced with the help of a pro-growth coalition which has its own financial interests at stake. What is new in all the cases studied, is the participation of a newer (at times more radical and aggressive but also more professional and better prepared) generation of environmental activists who make use of their own expertise, ranging from *technical* aspects – as in the case of the several involved lawyers or medical and toxicological experts – to more *humanistic* ones, such as in the case of the numerous activists who are also media studies experts, documentarists (e.g., Tsui Shu-hsin) and film-makers or musicians (Wu Zhi-ning); the latter employ their own knowledge in creative ways, crafting vivid pictures of all that would be lost if several of the developmental cases at stake were to be built, thereby easily making such plights widespread among a wider audience who is more easily attracted by graphic representations of environmental disasters than by lengthy legal caveats. The combination of these two elements has turned out to be a powerful weapon against the previously unbeatable dominance of the pro-development faction.

However, the developmental cases analyzed in the book clearly show that governance involves several other stakeholders than the government and that, until now, regulations, laws and legal deterrents have not been sufficient to prevent behaviors where the financial motivation is stronger than the risk of being discovered violating the law. In fact, while on the one hand economic dependency complicates the relationship between pollution and protest, with local residents often displaying mixed feelings towards industrial development and the improved condition that such projects might bring to a specific locality, there is a marked change between Taiwan of the 1980s – when people still valued economic and fiscal revenues above environmental concerns – and the past decade; thanks to social movements' and dedicated activists' efforts, and partly due to the numerous campaigns carried out by the government and EPA, Taiwanese' perception of environmental awareness has no doubt become much higher (Grano, 2014b).

Present-day industrial society is intertwined with technological progress and scientific advancements. Technology has no doubt greatly improved living standards around the world. A direct consequence though, has been the increase in pollution and ecological damages brought upon us by a proliferation of outdated technologies such as old engines and fossil fuels, to name but a few. Pollution caused by industrial wastes, power generation and motor vehicles are among today's leading sources of environmental problems (Yearley, 1991: 114). It is therefore no wonder that environmentalists are largely opposed to science and technology. Yet, in all of the cases I have studied in my book, as well as in the worst man-made ecological disasters that plague the world, the real culprit is the human race with its corruptible habits. In Taiwan even though *law in the books* is flawless, *law* de facto is often forgotten and a wide-spread laissez-faire attitude on part of the authorities has allowed corrupt practices to take root and continue undisturbed. In Kaohsiung, Taiwan's second biggest city after Taipei, a scandal that was under the media spotlight for a few weeks, unveiled a net of interests and dishonesty, to the detriment of environmental law enforcement (Chang, R., 2012: 3). The scam consisted in a conspiracy agreement between local environmental bureaus and university professors: the latter would serve as experts on committees for the review of relevant developmental cases, which would then be pushed forward without any serious assessment, catering solely to the interests of developers. These would then show their gratitude towards such complacent "reviewers" with financial rewards. At any rate, the mere fact that such a scandal caught the media attention illustrates that Taiwan's bureaucracy is no longer untouchable; activists have learned how to challenge governmental decisions by using those very same means for public participation that the government has created as valves to diffuse popular discontent.

While Taiwan has come a long way from the times in which the legal framework did not facilitate public challenges to administrative decisions, governmental agencies were still protected by various civil service and secrecy laws and environmental groups were unable to directly blame or target administrative units for their pro-development bias, the new constitutional asset nevertheless still presents some challenges for environmental activists and conservation experts. Thanks to an increase in the availability of suitable legal channels, the

Figure 7.1 Taipei 101 as a showcase for green technologies (source: André Beckershoff).

past decade has raised environmental groups' influence while their willingness to fight concentrated economic interests has grown; at the same time, though business interests have gained more influence and through the employment of those very same channels available to the public (e.g., the EIA process) they pay their own legal and professional experts to fight their game, in what has become a "battle of scientific opinions." Furthermore, the word "green" has by now become so mis-used and widespread that it often stands to represent those very same large-scale commercial interests (and products) that environmental groups are trying so hard to oppose. All politicians in Taiwan try to cultivate a green look by highlighting the positive changes they bring about for the environment such as developing the country's metro system and bicycle trails and establishing an effective recycling structure; therefore party politics and developmental plans that are deemed in the "national interest" are often sugar-coated in a "light green" discourse to give them an air of sanctity.

In Taiwan, where only big parties – with their better established electoral base – enjoy a substantial degree of influence, smaller parties (such as the Green Party) and other interest groups have had to learn how to navigate the murky waters of Taiwanese politics, by shifting their allegiances in tactical ways, as ecological concerns clashed with developmental ones. So while the Taipei City government is a loathed enemy in the Big Dome case, it becomes a precious ally in the Tamsui North Shore case, in which its interests merge and coincide with those of environmental groups, in a joint effort vis-à-vis the local government and the developers. The same "opportunistic" stance is devised on the part of the

Figure 7.2 Taipei 101 as a showcase for green technologies, August 29, 2011 (source: Simona Grano).

various governmental organs, so that green governance in Taiwan becomes a volatile process with ever-changing boundaries.

As mentioned at the beginning of this chapter, countries that have a poor record of environmental protection usually have a poor record in respecting human rights. Then there is a second type of country, which on the surface displays all the main traits of a democracy – by respecting its citizens' constitutional rights – but displays an utter lack of respect for its natural surroundings. This model fits Taiwan perfectly: a young – and still imperfect – democratic system in which endemic corruption hampers several of the good-willed efforts made by activists and concerned citizens to hinder environmentally unsustainable projects and practices. Underlining all this and often the real culprit for this disrespect of both nature and human health, sometimes referred to as "second-generation's rights,"[2] is the widespread corruption present at all levels of the bureaucratic system in Taiwan. The result is that numerous developmental projects are either pushed forth with no real concern for the damage to the environment, simply to please the developers and their financial interests, or said projects are constructed with shoddy materials; a situation that can have fatal consequences for both the people and the environment: the infamous video of a hotel collapsing under the rage of Typhoon Morakot in August 2009 was repeatedly shown in several news channels around the world.

In fact, corruption still affects the course of deliberations and public policy decisions in regards to environmental protection, especially when conservationists' issues come to clash with powerful pro-development interests. One of the biggest problems relating to any big developmental project is the priority accorded to economic profit over safety measures, which is a permanent feature of almost all developmental cases, from nuclear power plants and naphtha crackers to residential as well as infrastructure complexes. Furthermore, several projects are promoted without any real need for them; when they get stalled by public opposition, such cases turn into costly deals that will take decades to reach a conclusion, without possibly ever becoming operational; such is the case with NPP-4 so far and it is too soon to say whether a similar fate awaits the Tamsui North Shore project.

Finally, a few words about the most important piece of legislation among the rules-based participatory approaches available, the Environmental Impact Assessment. Even though the institute of the EIA has progressed from a closed-door discussion to an open review process, its claim of professionalism is often a mere lip-service to its name; in other words political influences still outweigh professionalism. This has been demonstrated in two of the cases analyzed, the Taipei Dome and the KPT petrochemical project in which, despite the solidity of the legal arguments collected by activists and their lawyers, the projects were either stopped due to political motivations (KPT) or went blatantly ahead in total disregard of the evidences collected by activists (Taipei Dome). Scholars (Ye, 1993: 26; Hsiao, 1999) as well as many of my sources among the academic and environmental communities, mentioned several times throughout the course of this book, have indicated that there are numerous problems still affecting the

EIA process, not least an overall lack of public confidence in this institute, which are in urgent need of fixing before Taiwan can achieve a truly modern and efficient environmental governance system.

Nevertheless, several changes for the better have taken place in the past few years. A positive variation from the past is the increased cooperation taking place among different stakeholders in the environmental camp. While, as seen in the first chapter, in the early days of Taiwan's environmental movement there was considerable tension between national-based environmental NGOs, which were predominantly middle-class organizations with intellectuals as their core members and conservation as their main agenda, and local activists working to protect their own rights, in all of the more recent case studies, social activists have worked side by side with local protesters, actively involving them in the choice of strategies to employ and assisting in various ways. Environmental groups have more chances to win against powerful corporations if they can mobilize sufficient support from the general public to aid their cause while channeling their grievances through political and legal channels, open to public scrutiny, such as in the cases of Kuokuang, Taipei Dome and Tamsui North Shore Road projects. In fact, the public responds to environmental issues not simply on the basis of facts but also of perceptions; these, in turn, can be affected by various elements and actors such as media, social activists and governmental actors. For these reasons the government, through its relevant agencies, needs to make sure that transparent and correct information reaches citizens, while at the same time it needs to educate the public about both the risks and the benefits of controversial facilities/projects. People have a right to know whether their urban surroundings and farms are polluted, and if so to what extent. They need this information to assess the risks they face and work out appropriate countermeasures. Transparency can also reduce public suspicions toward the government and keep protests from happening. In fact, not all protests on part of the environment are sensible. Some heavily criticized projects, such as the Kaohsiung or Taipei Metro System, initially came under fierce attack from the respective environmental communities even though their contribution to the lowering of urban smog is by now undisputed. Therefore, in order to quench the fear that often grips citizens opposing facilities in their neighborhoods, in a typical NIMBY syndrome fashion, the public has to be educated by the government on the costs and benefits of each method and/or project. Key to maintain public support is *transparency* in the planning and construction phase of such facilities, which is, at times, still insufficient. More transparency would in fact ensure the avoidance of situations in which activists a priori attack a project, without really being aware of its benefits/damages.

Disclosure is also important because policymakers at all levels need information to properly formulate preventive measures, identify polluters and impose penalties before environmental degradation gets worse. To paraphrase Williams and Chang (2008: 134) a successful environmental protection can only be achieved with direct participation and support on the part of citizens.

Finally, I have to admit that while in Taiwan a thing that never failed to impress me was the civility and respect for the "enemy" that activists displayed during their fights. Not once in the course of the numerous protests I have witnessed and taken part in have I seen violence, damages or property abuse. The opponent is verbally criticized, sometimes heavily, especially when it is a big powerful corporation, but in a civil way that we Europeans, accustomed to several "black-blocks style" of protest seen in the past two decades, in which everything is ransacked and broken, could benefit learning from. This attitude of civility was displayed also during the Sunflower Movement occupation of the Legislative Yuan when protesters and students exercised self-restraint and discipline in, for instance, keeping the grounds clean and tidy, an attitude that was frequently heralded, by sympathetic media outlets, as an example of extreme courtesy and civility. However, certain more radical behaviors, such as an anti-nuclear rally where activists (including Chen Wei-ting, one of the leaders of the Sunflower Movement) occupied Zhongxiao West Road, blocking the traffic for hours (Tsai and Chung, 2014: 3) have occurred, signaling that activists are beginning to resort to more radical methods of contention. In this regard, some scholars believe that the return to certain forms of dissent of the past, such as mass protests and unlawful occupation of public spaces, is a direct consequence of what protesters perceive as total disregard and indifference towards their pleas, on the part of the state. In Rawnsley's words:

> However, governmental indifference to popular sentiment is commonplace in all political systems. While breaking into and occupying the Legislative Yuan is illegal, such forms of protest are often a weapon wielded by the weak, an instrument of activism used by groups of people frustrated that their voices are not being heard through any other channel.... The existing democratic system, which is a legacy of the democratisation process of the 20th century, seems no longer adequate to serve the citizens of the 21st century. The island is in need of a "second wave" of democratisation.
>
> (Rawnsley, 2014a)

The Fukushima incident, one of the most effective events, which has contributed to relaunch nuclear awareness and environmental concerns, benefited from a favorable historical conjunction in which social movements, following the growing discontent against the Ma Ying-jeou administration since the KMT's return to power in 2008, were already beginning to resurface. It is not surprising that all of the protests analyzed took place in the post-2008 era, signaling that the KMT's comeback was a sudden wake-up call for social activists, who believe that many of the official channels that were previously available are closing one after the other as the government reaches important decisions behind closed-doors meetings. Protests thus became the only way through which concerned activists can make their requests heard (Fell, 2014). In Ho Ming-sho's words: "Thus, the Fukushima incident was for Taiwan less an out of the blue wake-up call from mass hibernation than a timely accelerator of the already activated civil society" (Ho, 2014a: 13).

This book looked at four developmental projects, with varying degrees of environmental damage starting from 2010 (three of the cases are ongoing controversies), in order to understand the trajectory of Taiwan's environmental politics, in light of the recent developments experienced by civil society. Even though the decisions to halt controversial projects were reached out of consideration for electoral campaigns, thus making political factors the most important in determining the outcome of a facility, the recent positive outcomes in obtaining the cancelation of Kuokuang, the stopping of NPP-4 as well as the legal victory in the case of Tamsui, highlight an unmistakable trend in which politicians' leadership of environmental campaign fades into the background while environmental activists gain strength and bipartisan support among the general populace, by mobilizing nationwide help and garnering professionals' aid.

The detailed analysis of the four case studies suggests that the political process in Taiwan is becoming more pluralistic in ways that are not yet sufficiently represented in the existing literature. The research questions raised in the introductory chapter, related to establishing which agents are more successful at influencing environmental regulations in loco and through which means, have been answered by showing how different cases have a specific outcome due to the actors involved and the methods used. In other words, environmental politics still take the back seat and activists' concerns are subjected to bigger issues (e.g., economic development) and political interests; however, when public opinion becomes involved, and cases become of national interest, then the government must take notice and yield to popular wishes. Thus, key factors affecting the positive or negative outcome in a fight against a particular project or policy change have much to do with both the nature of the policy (or project) as well as with the specific characteristics of the case and, in all the cases analyzed, the strategic timing of the protests, in a pre-electoral campaign year. In fact, both the decision to halt the KPT project, as well as to stop NPP-4 and the Tamsui Road Project were reached shortly before a big election (the presidential one in the case of KPT and the municipal elections of November 2014, in the case of Tamsui and NPP-4), thus showing that all politicians want to be perceived as "green" and seek to ingratiate the electorate in such sensitive moments.

Furthermore, differences and specificities such as the geography of a certain place or its physical attributes account, partly, for the maneuver space that activists need to launch their activities. For example, if a place is prone to seismic activity or if it is filled with cultural and archeological relics or if there are other developmental projects scheduled to be developed in that same area (as was the case with the first proposal to develop KPT in Yunlin, at the site of the sixth naphtha cracker), then the possibility that a variety of different actors (from the cultural relic bureau personnel, to the seismologic bureau to other external actors such as social groups), will try to interfere with the developmental case or influence its outcome, is much higher.

At any rate, Taiwan's environmental movement capacity has grown immensely in the past decade and has successfully outweighed the loss of a previous powerful political ally such as the DPP, while almost singlehandedly determining the outcome of many successful battles, sustained in the past few years. In fact, in all the cases analyzed, apart from their scientific knowledge and legal/administrative expertise, activists did not have other tools to oppose the pro-growth coalition; they thus sought to procrastinate the final decision, for as long as possible. Time is a precious commodity both for activists and developers, who have to undergo costly sanctions the longer a project is delayed and keep borrowing money from investors. The longer a decision is postponed the more chances activists have of winning. If a developmental conflict is protracted for too long, the pro-growth coalition will often dissolve under the weight of mounting economic debts and sanctions.

This book has sought to depict the trajectory of Taiwan's environmental governance approach in the past few years, comparing it with that of the past, trying to pinpoint its most important changes and transformations. The importance of this trajectory as a model for China to solve its environmental issues should not be underestimated since it is a known fact that, as Taiwan first embarked onto the path of democratization, its environmental movement played a crucial role. China is just about starting to recognize and tackle the costs of its economic success through a mixture of popular protests, public activism and governmental programs. Taiwan's experience, as amply shown in Chapter 2 of this book, is therefore of the utmost importance seen in this light. In fact the battle to protect the environment is, at the same time, a battle to further democratization.

Notes

1 Echo Lin is a long-time activist in both social as well as environmental movements and plays an important role in Amnesty International Taiwan.

2 First-generation rights include civil and political rights such as free speech and conscience and freedom from torture and arbitrary detention. Second-generation rights are social, economic and cultural and include the right to attain a reasonable level of education, healthcare and housing and the right to a clean and healthy environment.

References

Agoramoorthy, Govindasamy. 2009. "Enforcement Challenges of Taiwan's Wildlife Conservation and Animal Protection Laws," *Journal of International Wildlife Law and Policy*, 12(3): 190–209.

Apple Daily. 2011. Survey. May 1. At: www.appledaily.com.tw/appledaily/article/headline/20110501/33357000 (last accessed on October 15, 2012).

Apple Daily. 2014. "News Bulletin from the Songshan Tobacco Factory Tree Protection Site. Farglory Uses Force to Fell the Trees, Activists Use Their Bodies to Stop it, Climb up on Trees and Call for Help" [松菸護樹現場快報。 遠雄強勢砍樹 志工肉身阻擋上樹呼救 *Songshan hushu xianchang kuaibao – Yuanxiong qiangshi kanshu – zhigong roushen zudang shangshu hujiu*]. At: www.appledaily.com.tw/realtimenews/article/new/20140425/386245/ (last accessed on November 1, 2014).

Arrigo, Linda, previous spokesperson of the Green Party of Taiwan, several interviews on June 29, 30, July 1, 2, 20, November 3 and 26, 2011, Taipei; informal communication, July 20, 2011, Taipei; email communication on May 14, 2014.

Arrigo, Linda Gail and Gaia Puleston. 2006. "The Environmental Movement in Taiwan after 2000: Advances and Dilemmas," in Dafydd Fell, Henning Kloter and Chang Bi-Yu (eds), *What Has Changed? Taiwan Before and After the Change in Ruling Parties*, Research Unit on Taiwanese Culture and Literature, Ruhr University. Bochum: Vol. 4. Wiesbaden: Harassowitz Verlag, pp. 165–186.

Arrigo, Linda Gail, Tze-Luen Lin and Yvonne M. Lin. 1996. "Environmental Conditions and Environmental Law in Taiwan," in Antonella Capria *et al.* (eds), *A World Survey of Environmental Law*. Milan: Giuffrè Editore, pp. 765–777.

Arrigo, Linda Gail, Si Jilgilan and Si Maraos. 2000. "A Minority within a Minority: Cultural Survival on Taiwan's Orchid Island," *Cultural Survival Quarterly*, 45. Special Report on Taiwan.

Asia One News. 2011. "Over 2,000 Rally against Nuclear Plants in Taiwan." March 20. At: http://news.asiaone.com/News/Latest+News/Asia/Story/A1Story20110320-269104.html (last accessed on November 2, 2014).

Atomic Energy Council. 2006. *Radioactive Waste Management in Taiwan: Low Level Radioactive Waste Final Disposal*. At: www.aec.gov.tw/english/index.html (last accessed on February 12, 2015).

Atomic Energy Council. 2010. *The Technology Report of Opinion Poll of Emergency Plan on Nuclear Accidents in 2006–2009*. At: www.aec.gov.tw/english/index.html (last accessed on February 12, 2015).

Bain, Irene. 1993. *Agricultural Reform in Taiwan: From Here to Modernity?* Hong Kong: Chinese University Press.

Beck, Ulrich. 1986. *Risikogesellschaft – Auf dem Weg in eine andere Moderne.* Frankfurt, A.M.: Suhrkamp/Insel.
Benford, Robert D. and David A. Snow. 2000. "Framing Processes and Social Movements: An Overview and Assessment." *Annual Review of Sociology*, 26(1): 611–639.
Butler, D. 2011. "Nuclear Safety: Reactors, Residents and Risk." *Nature*, 472(7344): 400–401.
Central News Agency. 2011. "The Danshui Road Project Has Passed the EIA with a Few Conditions" [淡北道路專案環評有條件通過 *danbei daolu zhuan'an huanping you tiaojian tongguo*]. April 15. At: http://n.yam.com/cnabc/fn/201104/20110415752754.html (last accessed on November 3, 2014).
Chan Shun-kuei 詹順貴. Attorney at Law at Primordial Law Firm (元貞聯合法律事務所 *yuanzhen lianhe falü shiwusuo*), interview, November 18, 2011, Taipei; email communication, January 23, 2014.
Chang, Doris T. 2009. *Women's Movements in Twentieth-Century Taiwan*. Urbana: University of Illinois Press.
Chang, Peter Wushou 張武修. Professor of Environmental Health Science at Taipei Medical University and chairman of National Association for Radiation Protection, informal communication, August 31, 2011.
Chang, Peter Wushou, 2012. "Putting Nuclear Nightmares to Bed," *Taipei Times*. February 17: 8. At: www.taipeitimes.com/News/editorials/archives/2012/02/17/2003525658 (last accessed February 12, 2015).
Chang, Peter Wushou, Hwang Jau-Yuan and Joseph B.H. Chang. 2001. "Spread of 60 Contaminated Steel and its Consequences for Taiwan," *Health Physics*, 81(6): 655–660.
Chang, Rich. 2012. "More Raids for Kaohsiung Environmental Graft Probe," *Taipei Times*. August 8: 3. At: www.taipeitimes.com/News/taiwan/archives/2012/08/08/2003539751 (last accessed February 12, 2015).
Chang Sheng-hsiung 張勝雄. Professor of Transportation Management at Tamkang University, interview, September 24, 2011.
Chao Chia-wei. 2011. Member of the Executive Board of Green Citizen Action Alliance, interview, October 24, 2011.
Chao, Vincent. 2012. "Japan Disaster: Underwater Volcanoes Pose Risk to Plant, Activists Say," *Taipei Times*. March 15. At: www.taipeitimes.com/News/taiwan/archives/2011/03/15/2003498234 (last accessed on February 10, 2015).
Chen Chiao-hua 陳椒華. Former chairwoman of Taiwan Environmental Protection Union and current spokesperson of the Taiwan Water Conservation Alliance, interview, July 28, 2011, Kaohsiung.
Chen Chi-chung 陳吉仲. Professor at the Department of Applied Economics with a focus on agricultural economics at National Chung Hsing University (國立中興大學 *guoli zhongxing daxue*) and Secretary-General of the University since August 2011, email communication, August 25 and 31, 2011.
Chen Du-sheng. 2011. "Taiwan's Antinuclear Movement in the Wake of the Fukushima Disaster, Viewed from an STS Perspective," *East Asian Science, Technology and Society*, 5: 567–572.
Chen Fang-bin (pseudonym), local Zhu Wei resident who, for security reasons, prefers to be known by his nickname "Chocolate Daddy," interview, May 4, 2011, Tamsui.
Chen Hsin-yi. 2009a. "A Look Back at the Songshan Tobacco Factory," *Taiwan Panorama*, 8: 74–83.
Chen Hsin-yi. 2009b. "All For an Egg: A History of a Controversy," *Taiwan Panorama*, 8: 86–90.

Chen Ketty, informal conversation, June 19, 2014, London, School of Oriental and African Studies.

Chen Ketty and Michael Cole. 2013. "Wind Turbine Troubles," *Taipei Times*. June 17. At: www.taipeitimes.com/News/feat/archives/2013/06/17/2003564971/2 (last accessed February 10, 2015).

Chen Kobe. 2013. "Modernizing Matsu: From Fortress to Casino Resort," *Taiwan Panorama*, 38(6): 87–91.

Chen Lih-Horng and Hung-Chih Shih. 2010. "Land Use Planning in Taiwan: A History," in Roger Bristow (ed.), *Planning in Taiwan: Spatial Planning in the Twenty-First Century*, Routledge Research on Taiwan. London and New York: Routledge, pp. 19–50.

Chen Man-li 陳曼麗. Chairwoman of the Homemakers United Foundation, interview, July 20 and September 21, 2011, Taipei.

Chen, Melody, 2004. "Edgar Lin 'Pursuing a Colorful Life'," *Taipei Times*. December 20: 3. At: www.taipeitimes.com/News/taiwan/archives/2004/12/20/2003215939 (last accessed on August 27, 2014).

Chen Zhao-lun 陳昭倫. 2011. "Why Does the Pink Dolphin Need to Change its Path?" [白海豚為甚麼該轉變 *Baihaitun weishenmen gai zhuanbian?*] in Wu Sheng and Wu Ming-yi (eds), *Wetland, Petrochemical Industry and Island Images* [溼地, 石化, 島嶼的想像 *shidi, shihua, daoyu de xiangxiang*]. Taipei: Youlu Wenhua Publishing, pp. 91–93.

Chiu Hua-mei. 2012. "Rise of the Green Party Shows Need for Change," *Taipei Times*. January 31: 8. At: www.taipeitimes.com/News/editorials/archives/2012/01/31/2003524288 (last accessed on February 10, 2015).

Chiu Yu-Tzu. 2001. "Radioactive Rebar Linked to Cancer," *Taipei Times*. April 29: 2. At: www.taipeitimes.com/News/local/archives/2001/04/29/0000083627 (last accessed February 12, 2015).

Chou Chang-hung 周昌弘. Dean of the College of Life Sciences (生命科學院, *shengming kexueyuan*) of the Graduate Institute of Ecology and Evolutionary Biology of China Medical University (中國醫藥大學 *zhongguo yiyao daxue*) and Academia Sinica specialist in plant ecology and phytochemical ecology, interview, June 30, 2011, Taipei.

Chou Kuei Tien 周桂田. Professor at National Taiwan University's Graduate Institute of National Development, interview, October 14, 2011, Taipei.

Chow, Peter C.Y. 2011. "Globalization, Dynamic Comparative Advantage, and Taiwan's Drive for Sustainable Development," in Robert Ash, John W. Garver and Penelope B. Prime (eds), *Taiwan's Democracy: Economic and Political Challenges*. London: Routledge, pp. 166–192.

Chu Wan-wen. 2002. "Taiwan's Petrochemical Industry since Economic Liberalization," Research Fellow and deputy director at Sun Yat-Sen Institute for Social Sciences and Philosophy (ISSP). Academia Sinica Nankang.

Chu Yu-chi 朱雨其. Director general at EPA, of the Department of Environmental Monitoring and Information Management (環境監測及資訊處 *huanjing jiance ji zixunchu*). Interview with the author on November 2, 2011, Taipei.

Chung, Hui-yun. 2005. "Nuclear Dump Dispute on Orchid Island," ICE Case Study No. 159. At: http://www1.american.edu/ted/ice/orchid-waste.htm (last accessed on October 15, 2012).

Cole, Michael J. 2013. "Wind Power Firm Hires Thugs to Protect Site." *Taipei Times*. June 10. At: www.taipeitimes.com/News/taiwan/archives/2013/06/10/2003564438 (last accessed February 10, 2015).

Cole, Michael J. 2014. "Civic Activism and Protests in Taiwan: Why Size Doesn't (Always) Matter." Draft paper presented at the Conference on Social Movements in Taiwan at SOAS Centre of Taiwan Studies. June 16–18. Courtesy of the author.

Cole, Michael J. 2014. Speech at SOAS, June 16, 2014.

Dang, Winston. 2011. "Re-evaluate Kuokuang's Changhua County Plan." *Taipei Times*. January 28: 8. At: www.taipeitimes.com/News/editorials/archives/2011/01/28/2003494592 (last accessed on April 5, 2012).

Dang, Winston 陳重信. Ex Minister of Environmental Protection (2007–8) and Chair Professor at Taiwan Medical University. Informal communication, August 31, 2011, Taipei; interview, October 18, 2011, Taipei.

De Changy, Florence. 2012. "Taiwan Presses Ahead with Home-built Nuclear Power Plant Despite Safety Fears." *Guardian*. February 14. At: www.theguardian.com/world/2012/feb/14/taiwan-asia-pacific (last accessed on January 28, 2013).

Diamond, Larry. 1999. *Developing Democracy: Toward Consolidation*, Baltimore, MD: Johns Hopkins.

Enn, Rosa. 2012. "Indigenous Empowerement Through Collective Learning," *Multicultural Education & Technology Journal*, 6(3): 149–161.

Environmental Protection Administration, ROC (Taiwan). 2012. "Introduction to the Environmental Protection Administration," p. 4. At: www.epa.gov.tw/en/artshow.aspx?art=2007122413190511&path=9103&list=9044 (last accessed February 10, 2015).

Fang Jay. 2012. "Nuclear Safety Is a Contradiction," *Taipei Times*. September 8: 8. At: www.taipeitimes.com/News/editorials/archives/2012/09/08/2003542228 (last accessed February 12, 2015).

Fang Jay. 2013. "Nuclear Plant May Win Democracy," *Taipei Times*. August 5: 8. At: www.taipeitimes.com/News/editorials/archives/2013/08/05/2003568945 (last accessed February 12, 2015).

Fei Xiaotong, 1992. *From the Soil – The Foundation of Chinese Society* (A Translation of Fei Xiaotong's Xiangtu Zhongguo). Los Angeles: University of California Press.

Fell, Dafydd. 2011. "Election Season Returns to Taiwan: Prospects for Taiwan's National Elections in 2012," *Brookings Northeast Asia Commentary*, 47, March 8. At: www.brookings.edu/opinions/2011/0307_taiwan_election_fell.aspx (last accessed February 10, 2015).

Fell, Dafydd. 2012. *Government and Politics in Taiwan*. London: Routledge.

Fell, Dafydd. 2013. "The Nuclear Referendum Issue in Taiwan," *China Policy Institute Blog*. March 27. At: http://blogs.nottingham.ac.uk/chinapolicyinstitute/2013/03/27/the-nuclear-referendum-issue-in-taiwan/ (last accessed February 10, 2015).

Fell, Dafydd. 2014. "Importance of Social Movements in Taiwan," *Taipei Times*. March 20: 8. At: www.taipeitimes.com/News/editorials/archives/2014/03/20/2003586080/1 (last accessed February 12, 2015).

Fell, Dafydd, Henning Klöter and Bi-Yu Chang. 2006. *What Has Changed? Taiwan Before and After the Change in Ruling Parties*. Wiesbaden: Otto Harrassowitz Verlag.

Gereffi, Gary, 1990. "Paths of Industrialization: An Overview," in Gary Gereffi and David Wyman (eds), *Manufacturing Miracles*. Princeton: Princeton University Press, pp. 3–31.

Grano, Simona A. 2014a. "Perception of Risk Towards Nuclear in Taiwan and Hong Kong," *Taiwan in Comparative Perspective*, 5, July: 60–78. At: www.lse.ac.uk/asiaResearchCentre/countries/taiwan/TaiwanProgramme/Journal/JournalContents/TCP-5Grano.pdf (last accessed February 10, 2015).

Grano, Simona A. 2014b. "Change and Continuities: Taiwan's Post-2008 Environmental Policies," *Journal of Current Chinese Affairs*, 43(3): 129–159.

Grano, Simona Alba and Tu Ping-Lan. 2012. "Development vs. Environment in Taibei," *Journal of Current Chinese Affairs*, 2: 1–34.

Green Party Press Release. 2011. "200 Million for a Play? You Haven't Seen Anything, Wu Den-yih Squanders 15 Billion" [夢想家2億不夠看，吳敦義敗家150億！遠雄大巨蛋弊案記者會*Mengxiangjia 2 Yi bu gou kan, Wu Den-yi Baijia 150 Yi!*]. Press Conference on the Taipei Dome Farglory Scandal. November 14.

Hannigan, John A. 2006. *Environmental Sociology*. London: Routledge.

Harrison, Mark. 2012. "The Anti-Media Monopoly Movement in Taiwan," *The China Story*, December 20.

Her, Kelly. 2010. "Creating a Sustainable Homeland," *Taiwan Review*, 60(3): 4–11.

Ho Jung-chun, Kao Shu-fen, Wang Jung-der, Su Chien-tien, Patricia Lee Chiao-tzu, Chen Ruey-yu, Chang Hung-lun, Marco C.F. Leong and Peter Wushou Chang. 2013. "Risk Perception, Trust, and Factors Related to a Planned New Nuclear Power Plant in Taiwan after the 2011 Fukushima Disaster," *Journal of Radiological Protection*, 33(4): 773–789.

Ho Ming-sho. 2003. "The Politics of Anti-nuclear Protest in Taiwan: A Case of Party-dependent Movement (1980–2000)," *Modern Asian Studies*, 37(3): 683–708.

Ho Ming-sho. 2004. "Contested Governance between Politics and Professionalism in Taiwan," *Journal of Contemporary Asia*, 34(2): 238–253.

Ho Ming-sho. 2005a. "Weakened State and Social Movement: The Paradox of Taiwanese Environmental Politics After Power Transfer," *Journal of Contemporary China*, 14(43): 339–352.

Ho Ming-sho. 2005b. "Taiwan's State and Social Movements Under the DPP Government, 2000–2004," *Journal of East Asian Studies*, 5: 401–425.

Ho, Ming-sho. 2005c. "Protest as Community Revival: Folk Religion in a Taiwanese Anti-Pollution Movement," *African and Asian Studies*, 4(3): 237–269.

Ho Ming-sho. 2006. *Green Democracy: A Research on Taiwan's Environmental Movements* [綠色民主：台灣環境運動的研究 *Lüse Minzhu, Taiwan Huanjing yundong de yanjiu*]. Taibei: Qunxue Publishing House.

Ho Ming-sho. 2010a. "Understanding the Trajectory of Social Movements in Taiwan (1980–2010)," *Journal of Current Chinese Affairs*, 39(3): 3–22.

Ho Ming-sho. 2010b. "To Co-opt Social Ties: How Taiwanese Petrochemical Industry Neutralized Environmental Opposition," *Mobilization: An International Journal*, 15(4): 447–463.

Ho Ming-sho 何明修. Professor at the Department of Sociology at National Taiwan University, interview, August 30, 2011, Taipei; informal communication, November 24, 2011, Taiwan's National Library, Taipei.

Ho Ming-sho. 2012. "A Conflict in Environmental Cultures: Tea-serving Volunteers and Conservationists in Taiwan." Conference paper presented at Rikkyo University, June 30–July 1. Courtesy of the author.

Ho Ming-sho. 2013. "Lukang Anti-DuPont Movement (Taiwan)," in David A. Snow, Donatella della Porta, Bert Klandermans and Doug McAdam (eds), *The Wiley-Blackwell Encyclopedia of Social and Political Movements*. Volume II. Oxford: Wiley-Blackwell.

Ho Ming-sho. 2014a. "The Fukushima Effect: Explaining the Resurgence of the Anti-nuclear Movement in Taiwan," *Environmental Politics*, 23(6): 965–983.

Ho Ming-sho. 2014b. "Resisting Naphtha Crackers: A Historical Survey of Environmental Politics in Taiwan," *China Perspectives*, 3: 5–14.

Ho Ming-Sho and Shieh Jyh-Cherng 謝志誠. 2011. *The Journey of the Eighth Naphtha Cracker in Taiwan: The Story of Guoguang Shihua* [八輕遊台灣：國光石化的故事 *ba qing you taiwan: guoguang shihua de gushi*]. Taipei: Rive Gauche Publishing House.

Hoare, Michael Rand. 2013. *How the UK Learnt to Deal with Taiwan: Politics and Trade 1960–1980*. Talk held at Taiwan Studies Seminar, SOAS, February 20. Courtesy of the author.

Hong Sungook. 2011. "Where Is the Nuclear Nation Going? Hopes and Fears over Nuclear Energy in South Korea after the Fukushima Disaster," *East Asian Science, Technology, and Society*, 5(3): 409–415.

Hsiao, Michael Hsin Huang. 1999. "Environmental Movements in Taiwan," in Y.S. Lee and Y. So (eds), *Asia's Environmental Movements: Comparative Perspectives*. Armonk, NY and London: M.E. Sharpe, pp. 31–54.

Hsiao, Michael Hsin-Huang and C.H. Ku (eds). 2010. "Taiwan's Social Movements Are Getting Restarted" [台灣社會運動再出發 *Taiwan Shehui Yundong Zai Chufa*]. Taipei: Jiuliu.

Hsu, Jenny W. 2014. "Taiwan Government Agrees to Halt Construction on Nuclear Plant," *The Wall Street Journal*. April 27. At: http://online.wsj.com/news/articles/SB10001424052702304163604579527072873788850 (last accessed February 10, 2015).

Hsü Kuang-jung 徐光蓉. Professor of Atmospheric Sciences at Taiwan National University, Academic Committee Taiwan Environmental Protection Union, interview, 15 September, 2011.

Hsu Shu-hsiang. 2003. "Democratization, Intergovernmental Relations, and Watershed Management in Taiwan," *Journal of Environment and Development*, 12(4): 455–463.

Hsu, Stacy. 2014. "FDA Proposes Radiation Tests for Japanese Snacks," *Taipei Times*. October 30. At: www.taipeitimes.com/News/taiwan/archives/2014/10/30/2003603270 (last accessed February 10, 2015).

Hsueh Li-min, Hsu Chen-kuo and Dwight H. Perkins (eds). 2001. *Industrialization and the State: The Changing Role of the Taiwan Government in the Economy, 1945–1998*. Harvard Studies in International Development. Cambridge, MA: Harvard University Press.

Hua Chang-I. 2010. "Land Problems, Planning Failure, and the Pending National Land Planning Law," in Roger Bristow (ed.), *Planning in Taiwan: Spatial Planning in the Twenty-first Century*. London and New York: Routledge, pp. 51–73.

Huang, Gillan Chi-Lun, Tim Gray and Derek Bell. 2013. "Environmental Justice of Nuclear Waste Policy in Taiwan: Taipower, Government, and Local Community," *Development and Sustainability*, 15(6): 1555–1571.

Huang Zheng-qi 黃正琪. Lawyer at Chihlaw international business practice, email communication, August 11, 2014.

Hung Chin-fu 2014. "The Internet and Taiwan's New Civic Movement in the Information Age: Hung Chung-chiu's Case (2013)," *Asiascape: Digita Asia*, 1(1–2): 54–77.

Jacobs, Andrew. 2012. "Vote Holds Fate of Nuclear Power in Taiwan," *New York Times*. January 13. www.nytimes.com/2012/01/13/world/asia/nuclear-power-emerges-as-election-issue-in-taiwan.html?_r=1 (last accessed February 10, 2015).

Jacobs, Bruce J. 2012. *Democratizing Taiwan*. Leiden; Boston: Brill.

Jhan Jhang-cyuan 詹長權. 2011. "Looking at the Choice of Building or Not the Sixth Petrochemical Plant and *Guoguang Shihua* from Health Risks Posed by the Petrochemical Industry" [從石化業的健康危害, 看（六輕和國光石化去留）的抉擇 *cong shihuaye de jiangkang weihai, kan (liu qing he guoguang shihua quliu) de jueze*], in Wu Sheng and Wu Ming-yi (eds), *Wetland, Petrochemical Industry and Island Images* [溼地, 石化, 島嶼的想像 *shidi, shihua, daoyu de xiangxiang*]. Taipei: Youlu Wenhua Publishing, pp. 48–54.

Jobin, Paul. 2010. "Hazards and Protests in the 'Green Silicon Island'," *China Perspective*, 3: 46–62.

Jobin, Paul. Director, French Centre for Research on Contemporary China, CEFC, Taipei Office (2009–2013), interview, October 4, 2011, Taipei; email communication on October 24, 2012. Courtesy of the author.

Jobin, Paul. 2012. "The Socio-politics of Nuclear Safety in Taiwan after Fukushima Disaster," unpublished paper presented at the 1st World Congress of Taiwan Studies.

Jou, Sue-Ching. 2005. "Domestic Politics in Urban Image Creation: Xinyi as the 'Manhattan of Taipei'," in Reginald Yin-Wang Kwok (ed.), *Globalizing Taipei: The Political Economy of Spatial Development*. Oxon and New York: Routledge.

Kerr, George H. 1965. *Formosa Betrayed*. Boston, MA: Houghton Mifflin.

Kerr, George H. 1974. *Formosa: Licensed Revolution and the Home Rule Movement, 1895–1945*. Boulder, CO: Westview Press.

Khagram, Sanjeev and Saleem H. Ali. 2008. "Transnational Transformations – From Government-centric Interstate Regimes to Cross-sectoral Multi-level Networks of Global Governance," in Jacob Park, Ken Conca and Matthias Finger (eds), *The Crisis of Global Environmental Governance*. New York: Routledge, pp. 132–162.

Khanna, Tarun and Krishna G. Palepu. 1997. "Why Focused Strategies May Be Wrong for Emerging Markets," *Harvard Business Review*, 75(4): 41–51.

Ku, Helen, Hsiu-chuan Shih and Chris Wang. 2013. "Motions Passed to Cease Nuclear Plant Construction," *Taipei Times*. March 22: 1. At: www.taipeitimes.com/News/front/archives/2013/03/22/2003557678 (last accessed February 12, 2015).

Kuo Cheng-Tian. 2000. "Taiwan's Distorted Democracy in Comparative Perspective," *Journal of Asian and African Studies*, 35(1): 85–111.

Lai Fen-lan 賴芬蘭. Former secretary-general of the Green Party of Taiwan, informal communication, June 30, 2011 and July 3, 2011; interview, June 29, 2011, October 30, 2011 and November 24, 2011.

Lai Pin-yu 賴品瑀. 2013a. "Members of the Committee to Protect Trees Are in Favour of Maintaining the Hua Guang Old Trees, Overall Preservation Turns Out to Be Promising" [樹保委員支持 華光老樹全區保存現生機 *shubao weiyuan zhichi huaguang laoshu quanqu baocun xian shengji*]. July 17, *Taiwan Environmental Information Center*. At: http://e-info.org.tw/node/87351 (last accessed February 10, 2015).

Lai Pin-yu 賴品瑀. 2013b. "Taipei High Court Declares the EIA on the North Shore Road Project Invalid. New Taipei City: We Will Not Stop Construction. Green Groups: We Will Raise an Injunction" [北高行撤銷淡北道環評－新北市: 不停工 環團: 提假處分, *beigaoxing chexiao danbeidao huanping – xinbeishi: bu tinggong. Huantuan: tijia chufen*]. September 5, *Taiwan Environmental Information Center*. At: http://e-info.org.tw/node/90215 (last accessed February 10, 2015).

Lai Pin-yu 賴品瑀. 2014a. "Although the Rate of Gravel Trucks Entering at Night Has Increased Sixfold, the Songshan Tobacco Factory Egg Has Nevertheless Managed to Pass the Differential Analysis of Environmental Impact" [夜間砂石車出入將增6倍松菸大巨蛋環差闖關成功 *yejian shashiche churu jiang ceng 6 bei songyan dajudan huancha chuangguan chenggong*]. June 12, *Taiwan Environmental Information Center*. At: http://e-info.org.tw/node/99994 (last accessed on November 1, 2014).

Lai Pin-yu 賴品瑀. 2014b. Whether or Not a Referendum on NPP-4 Can Establish a Legal Precedent? "At the Legislative Hearing Both Sides Oppose Each Other with Harshness on the Pros and Cons" [核四公投能否成案？ 聽證會正反方針鋒相對 *Hesi gongtou nengfou cheng'an? Tingzhenghui zhengfan fang zhendui xiangdui*]. August 7, *Taiwan Environmental Information Center*. At: http://e-info.org.tw/node/101309 (last accessed on October 26, 2014).

Lai Pin-yu 賴品瑀. 2014c. "Just Moving Two Trees Has Turned into Just Leaving Two

Trees: The Harmonization Conference on Changing the Road Course Has Broken up in Discord" [只移兩棵樹變只留兩棵 大巨蛋路型變更協調會不歡而散 *zhi yi liang ke shu bian zhi liu liang ke shu – dajudan luxing bian geng xietiao hui bu huan er san*]. August 7, *Taiwan Environmental Information Center*. At: http://e-info.org.tw/node/101278 (last accessed on October 25, 2014).

Lai Pin-yu 賴品瑀. 2014d. "So that Trees and Stray Animals Will Have a Seat. Pan Hansheng Establishes the 'Tree Party' Together with 21 Young Individuals so as to Make a Break-through at the Next Big Election" [讓樹木、流浪動物有席次 潘翰聲等21青年組「樹黨」闖大選 *rang shumu, liulang dongwu you xici – Pan Han-qiang deng 21 qingnian zu (shu dang) chuang daxuan*]. September 5, *Taiwan Environmental Information Center*. At: http://e-info.org.tw/node/101941 (last accessed February 10, 2015).

Lai Rong-xiao 賴榮孝. Chairman of the Society of Wilderness, interview with the author, September 1, 2011; informal communication, September 1, 2011, Taipei.

Lan Wei-gong 藍維恭. Director of the Department of Engineering of the Ministry of Transportation, interview with the author and Tu Ping-lan, May 12, 2011, Tamsui; email communication, July 12, 2011.

Lee I-Chia. 2011a. "Anti-nuclear Rally Draws Legions," *Taipei Times*. May 1: 1. At: www.taipeitimes.com/News/front/archives/2011/05/01/2003502115 (last accessed February 12, 2015).

Lee I-Chia. 2011b. "Drop Taipei Dome Plan: Activists," *Taipei Times*. August 31: 2. At: www.taipeitimes.com/News/taiwan/archives/2011/08/31/2003512111 (last accessed on April 2, 2012).

Lee I-Chia. 2011c. "Make Tobacco Factory into a Park: Protesters," *Taipei Times*. October 31: 2. At: www.taipeitimes.com/News/taiwan/archives/2011/10/31/2003517124 (last accessed February 12, 2015).

Lee I-Chia. 2012a. "Group Raises Nuclear Awareness," *Taipei Times*. March 6: 2. At: www.taipeitimes.com/News/taiwan/archives/2012/03/06/2003527123 (last accessed February 10, 2015).

Lee I-Chia. 2012b. "Thousands March Against Nuclear Power," *Taipei Times*. March 12: 1. At: www.taipeitimes.com/News/front/archives/2012/03/12/2003527569 (last accessed February 10, 2015).

Lee I-Chia. 2012c. "Taipower Hits Back at Protesters," *Taipei Times*. March 13: 2. At: www.taipeitimes.com/News/taiwan/archives/2012/03/13/2003527677 (last accessed February 12, 2015).

Lee I-Chia. 2013a. "Science Park's Wastewater Policy Protested," *Taipei Times*. February 5: 3. At: www.taipeitimes.com/News/taiwan/archives/2013/02/05/2003554284 (last accessed February 10, 2015).

Lee I-Chia. 2013b. "Environmentalists Continue Sit-in Protest in Taipei," *Taipei Times*. April 1: 3. At: www.taipeitimes.com/News/taiwan/archives/2013/04/01/2003558523 (last accessed February 12, 2015).

Lee I-Chia. 2014a. "Do Not Extend Nuclear Plants' Life, Groups Urge," *Taipei Times*. February 16: 1. At: www.taipeitimes.com/News/front/archives/2014/02/16/2003583611 (last accessed February 10, 2015).

Lee I-Chia. 2014b. "Groups to Stage Anti-nuclear Run," *Taipei Times*. February 17: 3. At: www.taipeitimes.com/News/taiwan/archives/2014/02/17/2003583717 (last accessed February 10, 2015).

Lee I-Chia. 2014c. "Bands Perform for Free in Anti Nuclear Power Concert," *Taipei Times*. February 23: 3. At: www.taipeitimes.com/News/taiwan/archives/2014/02/23/2003584156 (last accessed February 12, 2015).

Lee I-Chia. 2014d. "Antinuclear Protest to Continue: Groups," *Taipei Times*. April 28: 1. At: www.taipeitimes.com/News/front/archives/2014/04/28/2003589062 (last accessed February 10, 2015).

Lee I-Chia. 2014e. "Activists File Corruption Petition," *Taipei Times*. June 7: 3. At: www.taipeitimes.com/News/taiwan/archives/2014/06/07/2003592194 (last accessed February 12, 2015).

Lee Su-Hoon. 1999. "Environmental Movements in South Korea," in Y.S. Lee and Y. So (eds), *Asia's Environmental Movements: Comparative Perspectives*, Armonk, NY and London: M.E. Sharpe, pp. 90–119.

Liao Jing-hui 廖靜蕙 2011. "The North Shore Road Project Is Approved with Conditions – Environmental Groups Intend to Sue" [淡北道路有條件通過－環團擬提告 *danbei daolu you tiaojian tongguo – huantuan ni tigao*]. April 18. *Taiwan Environmental Information Center*. At: http://e-info.org.tw/node/65724 (last accessed February 10, 2015).

Liberty Times. 2011. "President Ma Faces Crisis of Election," April 23.

Lin, Echo 林仁惠. Secretary-general of Environmental Jurists Association, interview, October 20, 2011, Taipei; email communication October 24, 2014.

Lin Le-hsin. 2010. "Erlin, Wanggong" [二林王功 *erlin wanggong*], in *Summer weeding – Anthology of Rural Grassroots Investigation Reports* [夏耘－農村草根調查文集 *xia yun– nongcun caogen diaocha wenji*]. Taipei: Taiwan Rural Front Publishing House.

Lin Le-hsin 林樂昕. Member of Taiwan Rural Front, interview, September 22 and 24, 2011, Taipei.

Lin, Rebecca. 2010. "The Kuokuang Petrochemical Controversy: Why Taiwan Land Is in Peril," *Commonwealth Magazine*, July 8, n. 450.

Lin, Sean. 2014. "Activists Protest Changhua Land Expropriation Plan," *Taipei Times*. July 22. At: www.taipeitimes.com/News/taiwan/archives/2014/07/22/2003595682 (last accessed February 10, 2015).

Lin Tze-luen, Alan 林子倫. Assistant professor, Department of Political Science, National Taiwan University, interview, July 15, 2011. Taipei.

Liu, Alan P.L. 1987. *Phoenix and the Lame Lion: Modernization in Taiwan and Mainland China, 1950–1980*. Stanford, CA: Hoover Institution.

Liu Li-er 劉黎兒. 2011. "We Cannot Afford a Single Nuclear Disaster: 52 Questions the Government Does Not Answer and Also Does Not Wish You to Know About" [我們經不起一次核災: 政府不回答, 也不希望你知道的 52 件事 *Women jing bu qi yi ci hezai: zhengfu bu huida, ye bu xiwang ni zhidao de 52 jian shi*]. Taipei: Xianjue Publishing House.

Lo, James Chi-hao. 2014. "Farglory To Be Punished if Dome Delayed," *Taipei Times*. June 6. At: www.chinapost.com.tw/taiwan/national/national-news/2014/06/06/409468/Farglory-to.htm (last accessed on October 11, 2014).

Loa Lok-sin. 2010a. "Activists Look To Buy Another 800 Hectares of Wetlands," *Taipei Times*. August 21: 2. At: www.taipeitimes.com/News/taiwan/print/2010/08/21/2003480947 (last accessed on October 11, 2014).

Loa Lok-sin. 2010b. "Group Calls for Purchases to Scupper Refineries Plan," *Taipei Times*. June 25: 2. At: www.taipeitimes.com/News/taiwan/archives/2010/06/25/2003476316 (last accessed on October 31, 2014).

Loa Lok-sin 2011. "Activists Urge Action on Attack," *Taipei Times*. March 25: 2. At: www.taipeitimes.com/News/taiwan/archives/2011/03/25/2003499076 (last accessed February 10, 2015).

Loa Lok-sin. 2012a. "Lanyu's Residents Grudgingly Accept Nuclear Storage," *Taipei Times*.

March 19: 3. At: www.taipeitimes.com/News/taiwan/archives/2012/03/19/2003528162 (last accessed on October 15, 2012).

Loa Lok-sin. 2012b. "Tao Protest Against Nuclear Waste," *Taipei Times*. March 2: 1. At: www.taipeitimes.com/News/front/archives/2012/03/02/2003526783 (last accessed on October 15, 2012).

Loa Lok-sin. 2012c. "Activists Demand Survey of Lanyu Radiation Levels," *Taipei Times*. September 29: 3. At: www.taipeitimes.com/News/taiwan/archives/2012/09/29/2003543947 (last accessed on October 15, 2012).

Loa Lok-sin. 2012d. "Tao Protest Against Nuclear Facility," *Taipei Times*. February 21: 1. At: www.taipeitimes.com/News/front/archives/2012/02/21/2003525985 (last accessed on October 15, 2012).

Loa Lok-sin. 2014. "Activists Say Construction Agency Fueling Speculation," *Taipei Times*. September 5: 3. At: www.taipeitimes.com/News/taiwan/archives/2014/09/05/2003599055 (last accessed February 12, 2015).

Loa Lok-sin and Shelley Shan. 2010. "Activists Warn EPA on Houli," *Taipei Times*. August 27: 2. At: www.taipeitimes.com/News/taiwan/archives/2010/08/27/200348141 7/2 (last accessed on November 3, 2014).

Lowther, William. 2014. "Sunflowers Altered Dynamics: Shriver," *Taipei Times*. June 29: 1. At: www.taipeitimes.com/News/front/archives/2014/06/29/2003593918 (last accessed February 10, 2015).

Lu Shi-wei 陸詩薇. Attorney at Law at the Wild at Heart Legal Defense Association, various interviews with the author, September 14, 16 and 19, 2011, Taipei; email communication, September 16, 2011.

Lu Shi-wei. 2013. *Formosa Plastics v. Prof. Tsuang Ben-Jei Update: Final Hearing of the First Instance on 29 August*. August 30. At: http://en.wildatheart.org.tw/story/321/7381 (last accessed February 10, 2015).

Lupke, Christopher. 2012. "Documenting Environmental Protest: Taiwan's Gongliao Fourth Nuclear Power Plant and the Cultural Politics of Dialogic Artifice," in Sylvia Li-chun Lin and Tze-Lan Deborah Sang (eds), *Documenting Taiwan on Film: Issues and Methods in New Documentaries*. London: Routledge, pp. 155–182.

Lyons, David. 2009. "The Two-headed Dragon: Environmental Policy and Progress Under Rising Democracy in Taiwan," *East Asia*, 26(1): 57–76.

Ma, Lydia. 2012. "Churches Join Masses in Telling President Ma, 'I'm a Man! I'm Antinuclear!'," World Mission – The Presbyterian Church in Taiwan (PCT). At: http://english.pct.org.tw/enNews_tcn.aspx?strBlockID=B00177&strContentid=C2012061200005&strCTID=&strDesc=Y&strPub=&strASP=enNews_tcn (last accessed on October 15, 2012).

Mathews, John and Jason Hu. 2013. "Renewable Energy vs. Nuclear Power: Taiwan's Energy Future in Light of Chinese, German and Japanese Experience since 3.11," *The Asia Pacific Journal*, June 10. At: http://japanfocus.org/-Mei_Chih-Hu/3954 (last accessed February 10, 2015).

McBeath, Gerald A. and Tse-Kang Leng. 2006. *Governance of Biodiversity Conservation in China and Taiwan*. Northampton, MA: Edward Elgar Publishing.

Mertha, Andrew. 2008. *China's Water Warriors: Citizen Action and Policy Change*. Ithaca, NY: Cornell University Press.

M'Gonigle, Michael and P. Ramsay. 2004. "Greening Environmental Law: From Sectoral Reform to Systemic Re-Formation," *Journal of Environmental Law and Practice*, 14: 333–356.

Ministry of Economic Affairs (MOEA). 1999. *Industrial Production Statistics Monthly, Taiwan Area*. Statistics Department, Taipei.

Mo Yan-chih. 2011a. "Taipei Dome Vote Stalls," *Taipei Times*. March 31: 2. www.taipeitimes.com/News/taiwan/archives/2011/03/31/2003499567 (last accessed February 12, 2015).

Mo Yan-chih. 2011b. "Water Shortage Growing," *Taipei Times*. May 11: 2. At: www.taipeitimes.com/News/taiwan/archives/2011/05/11/2003502957 (last accessed February 12, 2015).

Mo Yan-chih. 2013. "Hau's Nuclear Comments Spark Concern," *Taipei Times*. March 24: 3. At: www.taipeitimes.com/News/taiwan/archives/2013/03/24/2003557861 (last accessed February 12, 2015).

Molotch, Harvey. 1976. "The City as a Growth Machine: Toward a Political Economy of Place," *American Journal of Sociology*, 82: 309–330.

Molotch, Harvey. 1993. "The Political Economy of Growth Machines," *Journal of Urban Affairs*, 15: 29–53.

National Science Council. 2003. *Petrochemical Industries in Taiwan, R.O.C.* Taipei: Petrochemical Industry Association of Taiwan.

New York Times. 2014. "Taiwan's Gutter Oil Scandal," September 18. At: www.nytimes.com/2014/09/19/opinion/taiwans-gutter-oil-scandal.html?_r=0 (last accessed February 10, 2015).

Our Island [*Women de Dao* 我們的島]. 2011a. "Taiwan's Situation Previous to the Big Earthquake – Nuclear Crisis" [臺灣大震之前－核電危機 *Taiwan Da Zhen Zhiqian – Hedian Weiji*], n. 599, March 21.

Our Island [*Women de Dao* 我們的島]. 2011b. "Trees' Heaven? Or Trees' Graveyard?" [樹木天堂？樹木墳場？ *Shumu Tiantang? Shumu fenchang?*], n. 609, May 30.

Our Island [*Women de Dao* 我們的島]. 2011c. "NPP-4 – How Are You?" [核四你好嗎? *He si – Ni hao ma?*], n. 625, September 26.

Our Island [*Women de Dao* 我們的島]. 2012a. "Mazu's Gambling Hope" [馬祖賭希望 *Mazu Du Xiwang*], n. 642. February 13.

Our Island [*Women de Dao* 我們的島]. 2012b. "Radioactive Houses: 20 Years" [輻射屋二十年 *fushewu ershi nian*], n. 678. March 11.

Our Island [*Women de Dao* 我們的島]. 2012c. "The Truth That Cannot Be Bottled Up Any Longer" [栓不住的真相 *shuan bu zhu de zhenxiang*], n. 659, June 11.

Our Island [*Women de Dao* 我們的島]. 2012d. "How Much Radioactive Products Is it Safe to Eat?" [輻射食品吃多少？*Fushe Shipin Chi Duoshao*？], n. 671. September 3.

Our Island [*Women de Dao* 我們的島]. 2014. "When Road Meet Trees" [當路遇見樹], n. 756, May 12.

Pan Han-sheng 潘翰聲. Green Party of Taiwan former spokesperson. Founder of the Tree Party, interview, September 1, 2011 Taipei; informal communication, November 10, 2011. Green Party's office. Taipei; email communication, June 29, 2014.

Pan Han-sheng. 2011a. "Governments Don't Care About Trees," *Taipei Times*. July 14: 8. At: www.taipeitimes.com/News/editorials/archives/2011/07/14/2003508164 (last accessed February 12, 2015).

Pan Han-sheng. 2011b. "Big Dome in Deficit Asks for an Even Bigger Loan!" [大巨蛋虧錢還要貸－敗家敗更多, *Dajudan Kuiqian hai yao dai-baijia bai geng duo*!] blog. November 14. At: http://panhan3.pixnet.net/blog/post/36150469 (last accessed February 10, 2015).

Pan, Jason. 2013. "Environmentalists Hail 'Tambei Ruling'," *Taipei Times*. September 5: 1. At: www.taipeitimes.com/News/front/archives/2013/09/05/2003571385 (last accessed February 10, 2015).

Peopo Curiosity Baby@Taiwan. 2011. August 29. At: www.peopo.org/curiosity/post/89212 (last accessed on April 21, 2012).

Petition, 2008. "Online Petition Opposing the Construction of the Tamsui Road" [反淡北道路連署 *fan danbei daolu lianshu*]. At: http://campaign.tw-npo.org/sign.php?id=2008061305190300 (last accessed February 10, 2015).

PNN (PTS News Network). 2010. "It Is Good to Tell the Truth: Merits and Demerits of the Petrochemical Industry" [好好說實話：石化業功與過 *Hao Hao Shuo Shihua: Shihuaye Gong Yu Guo*]. October 11.

PNN (PTS News Network). 2012. "There Is Nobody Who Is Opposed to Nuclear Energy" [沒有人反對核能 *mei you ren fandui heneng*]. May 29. At: http://pnn.pts.org.tw/main/2012/05/29/沒有人反對核能 (last accessed February 10, 2015).

Power Technology. 2012. "Lungmen (Dragon Gate) Nuclear Project, Taiwan." At www.power-technology.com/projects/lungmen (last accessed on June 7, 2012).

Rawnsley, Ming-Yeh, 2014a. "New Civic Movements and Further Democratisation in Taiwan," *China Policy Institute Blog*. April 29. At: http://blogs.nottingham.ac.uk/chinapolicyinstitute/2014/04/29/new-civic-movements-and-further-democratisation-in-taiwan/ (last accessed February 10, 2015).

Rawnsley, Ming-Yeh. 2014b. "Anti-Media-Monopoly Policies and Further Democratisation in Taiwan," *Journal of Current Chinese Affairs*, 43(3): 105–128.

Reardon-Anderson, James. 1997. *Pollution, Politics, and Foreign Investment in Taiwan: The Lukang Rebellion*. Armonk, NY: M.E. Sharpe.

Rigger, Shelley. 1999. *Politics in Taiwan: Voting for Democracy*. London; New York: Routledge.

Rigger, Shelley. 2011. *Why Taiwan Matters: Small Island, Global Powerhouse*. Lanham, MD: Rowman & Littlefield.

Sabatier, Paul A. and Hank C. Jenkins-Smith. 1993. *Policy Change and Learning: An Advocacy Coalition Approach*. Boulder, CO: Westview Press.

"Save the Taiwan Pink Dolphins." 2012. At: http://taiwansousa.blogspot.com/ (last accessed on February 25, 2012).

Schafferer, Christian. 2001. "Taiwan's Nuclear Policy and Anti-nuclear Movement," in Christian Aspalter (ed.), *Understanding Modern Taiwan*. Aldershot: Ashgate, pp. 97–124.

Schubert, Gunther. 2012. "No Winds of Change: Taiwan's 2012 National Elections and the Post-Election Fallout," *Journal of Current Chinese Affairs*, 41(3): 143–161.

Shan, Shelley. 2011a. "Activists Vow Action if Work on Fourth Plant Continues," *Taipei Times*. March 16. At: www.taipeitimes.com/News/front/archives/2011/03/16/2003498300 (last accessed February 10, 2015).

Shan, Shelley. 2011b. "Japan's Disaster: Stop Building Fourth Plant," *Taipei Times*. March 18. At: www.taipeitimes.com/News/taiwan/archives/2011/03/18/2003498490/2 (last accessed February 10, 2015).

Shapiro, Don. 2002. "Rising to the Challenge," *Taiwan Review*, August: 24–31.

Shih Fang-Long. 2012. "Generating Power in Taiwan: Nuclear, Political and Religious Power," *Culture and Religion*, 13(3): 295–313.

Shih Hsiu-chuan. 2010. "Doctors Petition Against Kuokuang Naphtha Plant," *Taipei Times*. August 7: 2. At: www.taipeitimes.com/News/taiwan/archives/2010/08/07/2003479792/2 (last accessed February 10, 2015).

Shih Hsiu-chuan. 2013. "Referendum Referred for Second Reading," *Taipei Times*. 27 April: 1. At: www.taipeitimes.com/News/front/archives/2013/04/27/2003560817 (last accessed February 12, 2015).

Shih Hsiu-chuan. 2014. "Farglory Chairman Freed on Bail in Bribery Probe," *Taipei Times*. June 1: 1. At: www.taipeitimes.com/News/front/archives/2014/06/01/2003591720 (last accessed February 10, 2015).

Shih Yi-hsiang. 2013. "Taiwan Needs Civil Society to Rise Up," *Taipei Times*. June 29: 8. At: www.taipeitimes.com/News/editorials/archives/2013/06/29/2003565900 (last accessed February 12, 2015).

Shih, W.C. 2005. "Taiwan's International Environmental Policy: Balancing Trade and the Environment," in P.G. Harris (ed.), *Confronting Environmental Change in East and Southeast Asia: Eco-Politics, Foreign Policy, and Sustainable Development.* Tokyo: United Nations University Press, pp. 119–132.

Song Hong-yi 宋宏一. Member of the Society of Wilderness, interview with Simona Grano and Tu Ping-lan, March 17, 2011.

Songshan Tobacco Factory Park Union Press Release. 2011. Planking at the Big Dome – "Give Us Back the Green Spaces on our Stomach" [仆街一大巨蛋 還我綠地趴 *pujie dajudan – huan wo lüdi pa*]. October 2.

Stone, Clarence N. 1989. *Regime Politics: Governing Atlanta 1946–1988*. Lawrence: University Press of Kansas.

Su Chueh-Yu. 2006. "Cable-car Contractor's License Revoked Amid Controversy," *Taiwan News*. July 27: 2.

Sui, Cindy. 2011a. "Taiwan Endangered Species Focus of New Awareness," *BBC News*. February 1. At: www.bbc.co.uk/news/world-asia-pacific-12208493 (last accessed on February 16, 2012).

Sui, Cindy. 2011b. "Opposition to Taiwanese Chemical Plant Grows," *BBC News*. March 2. At: www.bbc.co.uk/news/12617084 (last accessed on February 26, 2012).

Sun Yu-Huay. 2013. "Taiwan Anti-Nuclear Protests May Derail $8.9 Billion Power Plant," *Bloomberg*. March 11. At: www.bloomberg.com/photo/taiwan-anti-nuclear-protests-may-derail-8-9-billion-power-plant-/-i2LsvG_AqTH4.html (last accessed on February 10, 2015).

Szasz, Andrew. 1999. "Grassroots Environmentalism in the United States: Implications for Asia's Environmental Movements," in Y.S. Lee and Y. So (eds), *Asia's Environmental Movements: Comparative Perspectives*. Armonk, NY and London: M.E. Sharpe, pp. 179–209.

Taipei Times. 2011. "Cracker May Be Built Overseas," January 25: 2. At: www.taipeitimes.com/News/taiwan/archives/2011/01/25/2003494383 (last accessed on October 15, 2014).

Taipei Times. 2013a. "Scrapping Nuclear Power Could Hurt GDP: Report," August 13: 14. At: www.taipeitimes.com/News/biz/archives/2013/08/13/2003569552 (last accessed on February 12, 2015).

Taipei Times. 2013b. "Taiwan Needs Energy Debate, German Diplomat Says," October 24: 4. At: www.taipeitimes.com/News/taiwan/archives/2013/10/24/2003575273 (last accessed on February 12, 2015).

Taipei Times. 2013c. "Taiwan and US Finalize Nuclear Fuel Agreement," December 22: 3. At: www.taipeitimes.com/News/taiwan/archives/2013/12/22/2003579563(last accessed on February 12, 2015).

Taiwan Environmental Information Center. 2010. 台灣回顧: 六輕18天內兩場大火 雲林鄉親圍廠抗議. At: http://e-info.org.tw/node/62640 (last accessed February 10, 2015).

Taiwan Insights. 2013. "Amid Massive Anti-nuclear Protests, Taiwanese Rethink Their Desired Lifestyle," March 29. At: www.taiwaninsights.com/tag/green-citizens-action-alliance/ (last accessed February 10, 2015).

Taiwan Rural Front. 2010. *Summer Weeding – Anthology of Rural Grassroots Investigation Reports* [夏耘－農村草根調查文集 *Xia Yun-Nongcun Caogen Diaocha Wenji*]. Taipei: Taiwan Rural Front.

Taiwan Statistical Data Book. 1996. Taipei: Council of Economic Planning and Development, p. 94.

Tang Chia-ling. 2011. "DPP Dismisses Ma's Opposition to Kuokuang Complex as Electioneering," *Taipei Times*. April 24: 3. At: www.taipeitimes.com/News/taiwan/archives/2011/04/24/2003501564 (last accessed on April 5, 2012).

Tang Shui-yan and Tang Ching-ping. 1997. "Democratization and Environmental Politics in Taiwan," *Asian Survey*, 37(3): 281–294.

Tang Shui-yan and Tang Ching-ping. 1999. "Democratization and the Environment: Entrepreneurial Politics and Interest Representation in Taiwan," *The China Quarterly*, 158: 350–366.

Tang Shui-yan and Tang Ching-ping. 2004. "Local Governance and Environmental Conservation: Gravel Politics and the Preservation of an Endangered Bird Species in Taiwan," *Environment and Planning*, 36(1): 173–189.

Tang Ching-ping. 2001. "Democratic Governance and Environmental Protection" [民主治理與環境保護 *minzhu zhili yu huangjing baohu*], *Taiwanese Political Science Journal*, 5: 178–217.

Tang Ching-ping. 2003. "Democratizing Urban Politics and Civic Environmentalism in Taiwan", *The China Quarterly*, 176: 1029–1051.

The China Post. 2011a. "Gov't to Discuss Anti-drought Measures," May 10. At: www.chinapost.com.tw/taiwan/national/national-news/2011/05/10/301758/Govt-to.htm (last accessed February 10, 2015).

The China Post. 2011b. "Ma Calls for Long-Term Plan to Fight Droughts, Floods," May 12. At: www.chinapost.com.tw/taiwan/national/national-news/2011/05/12/302050/Ma-calls.htm (last accessed February 10, 2015).

The China Post. 2011c. "Building of Taipei Dome to Begin This Year," June 17. At: www.chinapost.com.tw/taiwan/local/taipei/2011/06/17/306530/Building-of.htm (last accessed on April 2, 2012).

The Economist. 2011. "Plastic Unfantastic," June 16. At: www.economist.com/node/18837149 (last accessed on January 10, 2013).

Tomlinson, Simon. 2013. "Mass Fight in Taiwanese Parliament over Debate on Nuclear Power," *Daily Mail*. August 2. At: www.dailymail.co.uk/news/article-2383620/Mass-fight-Taiwanese-parliament-debate-nuclear-power.html?ito=feeds-newsxml (last accessed February 10, 2015).

Tsai Chang-sheng and Jake Chung. 2014. "Zhongxiao W Road Occupation Sparks Huge Online Debate," *Taipei Times*. April 29: 3. At: www.taipeitimes.com/News/taiwan/archives/2014/04/29/2003589162 (last accessed February 10, 2015).

Tsai, June. 2011. "Safety of Taiwan's 4th Nuclear Power Plant Called into Question," *Taiwan Today*. December 20. At: www.taiwantoday.tw/ct.asp?xItem=182684&ctNode=445 (last accessed on October 15, 2012).

Tsai, June. 2012. "Nuclear Concerns Dominate Taiwan Environmental Poll," *Taiwan Today*. January 2. At: www.taiwantoday.tw/ct.asp?xItem=183568&ctNode=445 (last accessed on October 15, 2012).

Tsai Pei-hui Frida 蔡培慧. Associate professor at Shih Hsin University's Graduate Institute for Social Transformation Studies, Taiwan Rural Front Spokesperson, interview, September 22, 2011; informal communication at Taiwan's National University Farmers' Market, October 15, 2011, Taipei and at the Green Party's Weekly Study Group Session, November 3, 2011, Taipei.

Tsai Ya-ying 蔡雅瀅. Lawyer at Wild at Heart Legal Foundation, interview, October 30, 2011; personal email communication, June 20, 2014.

Tsai Ying-wen 蔡英文. 2011. "2025 Plans for a Nuclear Free Home" [2025非核家園計畫 2025 *feihe jiayuan jihua*]. September 2. At: www.iing.tw/p/2025.html (last accessed on October 15, 2012).

Tseng Te-jung, Mei-hsueh Ling and Stacey Hsu. 2013. "Actor Joseph Cheng Makes First Anti-Nuclear Film," *Taipei Times*. August 11: 3. At: www.taipeitimes.com/News/taiwan/archives/2013/08/11/2003569435 (last accessed February 12, 2015).

Tsuang Ben-jei 莊秉潔. Professor at National Chung-hsing University, Department of Environmental Engineering, email communication, November 12, 2012.

Tsui Shu-Hsin 崔愫欣. 2004. "Gongliao, How Are You?" [貢寮你好嗎?]. Documentary.

Tsui Shu-Hsin 崔愫欣. 2009. "The Crisis of the Construction of the Road Has Not Passed Yet – Tamsui Riverbank Is at Stake" [蓋路危機仍未解除-淡水河岸岌岌可危 *Gailu weiji reng wei jiechu – danshui he'an jijikewei*]. At: www.gcaa.org.tw/modules/tadnews/index.php?nsn=91 (last accessed on June 16, 2013).

Tsui Shu-hsin 崔愫欣. Secretary-General of the Green Citizens' Action Alliance and member of the *NNP4 Safety Committee*, various interviews: March 24, 2011, September 8, 2011, September 17 and September 20, 2011 and informal conversation at Green Party Study Group on September 22, 2011; email communication July 6, 2012 and March 22, 2014.

Tu Ying-yi. 2001. "The Petrochemical Industry," in Li-min Hsue, Chen-kuo Hsu and Dwight H. Perkins (eds), *Industrialization and the State: The Changing Role of the Taiwan Government in the Economy 1945–1998.* Harvard Studies in International Development. Cambridge, MA: Harvard University Press, pp. 225–265.

Ungar, Sheldon. 1992. "The Rise and (Relative) Decline of Global Warming as a Social Problem," *The Sociological Quarterly*, 33(4): 483–501.

United Press International. 2013. "Taiwan Nuclear Power Plant Progresses amid Resistance," September 11. At: www.upi.com/Business_News/Energy-Resources/2013/09/11/Taiwan-nuclear-power-plant-progresses-amid-resistance/UPI-29971378927865/#ixzz2jlswQkjO (last accessed on January 28, 2013).

Vaughan, Elaine and Marianne Seifert. 1992. "Variability in the Framing of Risk Issues," *Journal of Social Issues*, 48(4): 119–135.

Voyer, Boris. Former employee of Taiwan's Ministry of Foreign affairs, telephone interview, November 16, 2011.

Wang, Audrey. 2010. "MOE Backs Kuokuang Petrochemical Plant," *Taiwan Today*. July 22. At: www.taiwantoday.tw/ct.asp?xItem=111764&ctNode=413 (last accessed on January 28, 2013).

Wang Chung-Ming 王鐘銘, Green Party Candidate for the position of City Councillor in Tamsui Town in 2010 (current Green Party spokesperson), various interviews, May 4, 2011 (together with Tu Ping-lan); June 29, 2011; August 1, 2011 and informal communication on September 17, 2011; email communication, October 12, 2013; Facebook post, August 13, 2014.

Wang, Chris. 2013a. "Thousands Protest Against Ma," *Taipei Times*. January 14. At: www.taipeitimes.com/News/front/archives/2013/01/14/2003552475 (last accessed on January 28, 2014).

Wang, Chris. 2013b. "Taipower Rate Hike Violates Resolution, DPP Says," *Taipei Times*. August 27. At: www.taipeitimes.com/News/taiwan/archives/2013/08/27/2003570693 (last accessed on January 28, 2014).

Wang, Chris. 2014. "Public Shows its Support for Student-led Protest," *Taipei Times*. March 31: 1. At: www.taipeitimes.com/News/front/archives/2014/03/31/2003586929 (last accessed February 10, 2015).

Wang, Chris and Lee I-Chia. 2013a. "Taipower's Stress Test Fraudulent: DPP," *Taipei Times*. May 8: 3. At: www.taipeitimes.com/News/taiwan/archives/2013/05/08/2003561747 (last accessed on January 28, 2013).

Wang, Chris and Lee I-Chia. 2013b. "Activists Say Plan a Risk to the Environment," *Taipei Times*. October 2: 4. At: www.taipeitimes.com/News/taiwan/archives/2013/10/02/2003573515 (last accessed February 10, 2015).

Weller, Robert P. 1999. *Alternative Civilities: Democracy and Culture in China and Taiwan*. Boulder, CO: Westview.

Weller, Robert P. 2006. *Discovering Nature: Globalization and Environmental Culture in China and Taiwan*. Cambridge: Cambridge University Press.

Wen Lii. 2014. "Elections 2014: Smaller Parties Clinch Big Victories," *Taipei Times*. December 1: 3. At: www.taipeitimes.com/News/taiwan/archives/2014/12/01/2003605715 (last accessed February 10, 2015).

Whitman, Darrell. 2008. "Stakeholders and the Politics of Environmental Policymaking," in Jacob Park, Ken Conca and Matthias Finger (eds), *The Crisis of Global Environmental Governance*. New York: Routledge, pp. 163–192.

Williams, Jack. 1992. "Environmentalism in Taiwan," in Denis Fred Simon and Michael Y.M. Kau (eds), *Taiwan: Beyond the Economic Miracle*. Armonk, NY: M.E. Sharpe, pp. 187–210.

Williams, Jack F. and David Ch'ang-yi Chang. 2008. *Taiwan's Environmental Struggle: Toward a Green Silicon Island*. New York: Routledge.

Winkler, Robyn. Attorney at Law at Wild at Heart Legal Defense Association, interview, September 5, 2011.

World Nuclear Association. 2014. "Nuclear Power in Taiwan." At: http://world-nuclear.org/info/inf115_taiwan.html (last accessed on February 12, 2015).

WPO. 2008. "World Publics Strongly Favor Requiring More Wind and Solar Energy, More Efficiency, Even if it Increases Costs." At: www.worldpublicopinion.org/pipa/pdf/nov08/WPO_Energy_Nov08_longart.pdf (last accessed February 10, 2015).

Wu Je-ming, 1990. *Social Protests During the Period of Transformation of Taiwan's Political Regime*. (In Chinese.) Master Thesis, Graduate Institute of Political Science, National Taiwan University.

Wu Kuen-yuh 吳焜裕. Associate Investigator for the Division of Environmental Health and Occupational Medicine at the National Taiwan University's Institute of Occupational Medicine and Industrial Hygiene. Member of Taiwan Watch, in charge of the Health Risk Assessment in the KPT case, interview, September 20, 2011.

Wu Sheng 吳晟. Writer and poet belonging to the *Bentuhua* current, interview, October 15–16, 2011.

Wu Sheng. 2011. "The Eight's Petrochemical Plant's Journey Around Taiwan: Reflections on the Book" [八輕遊台灣 – 讀後感], in Ho Ming-Sho and Shieh Jyh-Cherng (eds), *The Eight's Petrochemical Plant's Journey Around Taiwan: The story of Guoguang Petrochemical* [八輕遊台灣：國光石化的故事 *ba qing you taiwan: guoguang shihua de gushi*]. Taipei: Rive Gauche Publishing House, pp. 4–9.

Wu Sheng and Wu Ming-yi 吳明益. 2011. *Wetlands, Petrochemical Industry, Island Images* [溼地・石化・島嶼想像, *Shidi, Shihua, Daoyu Xiangxiang*]. Taipei: Unique Route Publishing House.

Xin Yage. 2013. "Agreement Between Government and Residents Over Expropriated Miaoli Land," *Asianews*. August 20. At: www.asianews.it/news-en/Agreement-between-government-and-residents-over-expropriated-Miaoli-land-28778.html (last accessed February 10, 2015).

Xu Xin-xin 許心欣. 2011. "Women, Environmental Protection and Economy" [女性。環保。經濟 *nuxing, huanbao, jingji*]. *Homemakers United Foundation Magazine*, 245, June: 8–9.

Yang Zijie. 2011. "Innovation in Anti-Nuke protest." At: http://taiwannonuke.blogspot.com/ (last accessed on July 1, 2011).

Ye Jun-rong. 1993. "Public Participation in Environmental Impact Assessment: The Requirements of Legal Norms and Practical Considerations" [环境影响评估的公共参与- 法规范的要求与现实的考虑 *Huanjing yingxiang pinggu de gonggong canyu – fagui fan de yaoqiu yu xianshi de kaolu*], *Society and Law Collection*, 11, pp. 17–42.

Yearley, Steven. 1991. *The Green Case: A Sociology of Environmental Issues, Arguments and Politics*. London: HarperCollins Academic.

Yo, Arthur 游藝. Resident of Xinyi District and director of the Songshan Tobacco Factory Park Union, interview, October 21 and 30, 2011; email communication, June 29, July 4, 2014.

Yuen, Samson. 2014. "Under the Shadow of China," *China Perspectives*, 2: 69–76.

Zhang Yu-yin 張譽尹. Executive director of the Environmental Jurists Association and lawyer at the Lin and Shi legal firm (永信法律事務所), email communication, August 10 and August 15, 2014.

Zheng Yongnian and Joseph Fewsmith. 2008. *China's Opening Society*. Oxon: Routledge.

Zhong Sheng-Xiong 鐘聖雄. 2011. "Oppose Taipei Big Egg and Ask for Green Spaces – Environmental Group Fends off Rain with Planking Protest" [反巨蛋要綠地 環團冒雨仆街抗議 *Fan Judan Yao Lvdi Huantuan Maoyu Pujie Kangyi*]. *PTS News Network*. October 2. At: http://pnn.pts.org.tw/main/?p=32988 (last accessed on January 28, 2013).

Zhu Jieming. 1999. "Local Growth Coalition: The Context and Implication of China's Gradualist Urban Land Reforms," *International Journal of Urban and Regional Research*, 23(3): 534–548.

Laws and regulations

Act on Sites for Establishment of Low Level Radioactive Waste Final Disposal Facility (低放射性廢棄物最終處置設施場址設置條例 *di fangshexing feiqiwu zuizhong chuzhi sheshi chang zhi shezhi tiaoli*) 2006. Taipei: Atomic Energy Council, Executive Yuan, http://erss.aec.gov.tw/law/LAWContentDetails.aspx?id=FL039264&KeyWordHL=StyleType=1 (last accessed on November 3, 2014).

Administrative Procedure Act (行政程序法 *xingzheng chengxu fa*), promulgated in 1999 and revised in 2005. At: http://law.moj.gov.tw/LawClass/LawAll.aspx?PCode=A0030055 (last accessed on November 3, 2014).

Assembly and Parade Act (集會遊行法 *jihui youxing fa*), promulgated in 1988 and revised in 1992 and 2002. At: http://law.moj.gov.tw/Eng/LawClass/LawHistory.aspx?PCode=D0080058 (last accessed on November 3, 2014).

Cultural Heritage Preservation Act (中華民國文化資產保存法 *zhonghua minguo wenhua zichan baocun*), promulgated in 1982 and revised in 2005. At: www.cca.gov.tw/law.do?method=find&id=30 (last accessed on April 22, 2012).

Environmental Impact Assessment Act (環境影響評估法 *huanjing yingxiang pingu fa*) promulgated in 1994 and revised in 1999 and in 2003. At: http://law.epa.gov.tw/zh-tw/laws/379692190.html#ch01 (last accessed on April 16, 2012).

Environmental Protection Law (中華民國環境保護法 *zhonghua minguo huanjing baohu fa*). At: http://law.epa.gov.tw/en/laws/506748469.html (last accessed on November 3, 2014).

Highway Law (中華民國公路法 *zhonghua minguo gonglu fa*), promulgated in 1959 and revised in 2010. At: http://law.moj.gov.tw/LawClass/LawAll.aspx?Pcode=K0040001 (last accessed on June 18, 2011).
National Preservation Project of Taiwan's Shore and Tidal Lands (臺灣沿海地區自然環境保護計畫 *taiwan yanhai diqu ziran huanjing baohu jihua*). At: www.cpami.gov.tw/chinese/index.php?option=com_content&view=article&id=10714&Itemid=53 (last accessed on November 3, 2014).
Non-Urban Land Use Regulation (非都市土地使用管制規則 *fei dushi tudi shiyong guanzhi guize*) promulgated in 1976 and revised in 2004 and 2011. At: www.land.moi.gov.tw/law/chhtml/newlawdetail.asp?nid=1238 (last accessed on November 3, 2014).
Referendum Act (公民投票法 *gongmin toupiao fa*) promulgated in 2003 and amended in 2009. At: http://law.moj.gov.tw/LawClass/LawAll.aspx?PCode=D0020050 (last accessed on November 3, 2014).
Standards for Determining Specific Items and Scope of Development Activities Requiring Environmental Impact Assessments (開發行為環境影響評估作業準則 開發行為應實施環境影響評估細目及範圍認定標準 *kaifa xingwei ying shishi huanjing yingxiang pinggu jimu ji fanwei rending biaozhun*) promulgated in 2009 and revised in 2013. At: http://ivy5.epa.gov.tw/epalaw/search/LordiDispFull.aspx?ltype=03&lname=0050 (last accessed on November 3, 2014).
Urban Planning Law (都市計畫法 *dushi jihua fa*) promulgated in 2006 and revised in 2010. At::http://law.moj.gov.tw/LawClass/LawContent.aspx?pcode=D0070001 (last accessed on November 3, 2014).
Water Resources Law (中華民國水利法 *zhonghua minguo shuili fa*), promulgated in 1942 and revised the last time in 2008. At: www12.water.gov.tw/eng/index.php?act=detail&ID=333 (last accessed on June 18, 2011).
Wetland Conservation Law (中華民國溼地保育法 *zhonghua minguo shidi baoyu fa*) enacted March 7, 2013. At: www.cpami.gov.tw/chinese/index.php?option=com_content&view=article&id=16504&Itemid=57 (last accessed on November 3, 2014).
Wildlife Conservation Law (中華民國野生動物保育法 *zhonghua minguo yesheng dongwu baoyu fa*) enacted on June 23, 1989 and amended in 1994, 2002, 2004, 2006 and 2013. At: http://law.moj.gov.tw/LawClass/LawContent.aspx?pcode=M0120001 (last accessed on November 3, 2014)

Useful websites

Environmental Protection Administration, Taiwan (EPA) www.epa.gov.tw.
Environmental Jurists Association (EJA) www.eja.org.tw/2510520497303402021921209.html.
Economic Cooperation Framework Agreement (ECFA) www.ecfa.org.tw.
Homemakers United Foundation (HUF) www.huf.org.tw/en/about.
Society of Wilderness (SOW) http://sow.tw/english/.
Taiwan Environmental Protection Union (TEPU) www.tepu.org.tw.
Taiwan Environmental Information Center (TEIA) http://e-info.org.tw.

Index

Page numbers in **bold** denote figures.

For Product Safety Concerns and Information please contact our EU representative GPSR@taylorandfrancis.com Taylor & Francis Verlag GmbH, Kaufingerstraße 24, 80331 München, Germany